LOW-FREQUENCY WAVES AND IRREGULARITIES
IN THE IONOSPHERE

ASTROPHYSICS AND SPACE SCIENCE LIBRARY

A SERIES OF BOOKS ON THE RECENT DEVELOPMENTS

OF SPACE SCIENCE AND OF GENERAL GEOPHYSICS AND ASTROPHYSICS

PUBLISHED IN CONNECTION WITH THE JOURNAL

SPACE SCIENCE REVIEWS

VOLUME 14

LOW-FREQUENCY WAVES AND IRREGULARITIES IN THE IONOSPHERE

PROCEEDINGS OF THE 2ND ESRIN-ESLAB SYMPOSIUM,
HELD IN FRASCATI, ITALY, 23–27 SEPTEMBER, 1968

Edited by

N. D'ANGELO

European Space Research Institute, Frascati (Rome), Italy

SPRINGER-SCIENCE+BUSINESS MEDIA, B.V.

The symposium was jointly sponsored by
the European Space Research Institute (Frascati, Italy) and
the European Space Laboratory (Noordwijkerhout, The Netherlands) of
the European Space Research Organisation (ESRO)

ISBN 978-94-010-3404-3   ISBN 978-94-010-3402-9 (eBook)
DOI 10.1007/978-94-010-3402-9

FOREWORD

During the last week of September 1968, ESRIN (the European Space Research Institute) held the ESRIN-ESLAB Symposium on 'Low-Frequency Waves and Irregularities in the Ionosphere' in Frascati, near Rome. The symposium was attended by about 60 participants, including speakers from most of the ESRO member states, the U.S.A., the U.S.S.R., and Peru.

The main topics covered were: (a) observations of ionospheric irregularities by radar scattering, (b) scintillations of satellite signals, (c) geomagnetic micropulsations, and (d) whistlers. Both theoretical and observational aspects were treated. In addition, laboratory results on low-frequency waves in plasmas were discussed, emphasis being given to their possible relevance to low-frequency ionospheric phenomena. Finally, a brief presentation (not included in these proceedings) of the ESRO rocket and satellite program was given by Dr. Pedersen of ESLAB.

The symposium provided an exchange of information among workers in closely related fields. It was also valuable in bringing together people whose experience is predominantly in ionospheric observations with others whose field of interest is mainly in plasma physics (theoretical or laboratory) – a combination that seemed particularly appropriate to ESRIN's program and functions.

Several ESRIN staff members were instrumental in the organization of the meeting; among them Dr. G. Fiocco and Dr. K. Schindler, who helped in defining the scientific program. It is a pleasure to thank Miss M. Sachs, who did all the real work both in the preparation of the conference and for the publication of its proceedings. Permission from the copyright holders to reproduce some of the papers and a number of figures in these proceedings is gratefully acknowledged.

N. D'ANGELO

Frascati, February 1969

TABLE OF CONTENTS

INTRODUCTORY LECTURE ON ION WAVES

DIETER PFIRSCH

Max-Planck-Institut für Astrophysik, München, W. Germany

This lecture is to give to some extent a background for the understanding of the observations of low-frequency waves in the ionosphere.

Certainly one is dealing with some kind of plasma oscillations. What can be their nature? There are quite a few possibilities for describing a plasma: one can regard it as an electrically conducting fluid, this being the so-called one-fluid model or magnetohydrodynamic approximation; or one can treat it as being composed of an electron fluid interpenetrated by an ion fluid (the two-fluid model). Both these models imply, as you know from ordinary hydrodynamics, that the collision mean free path of the particles is small compared with the macroscopic scale lengths involved. For our purposes these are the wavelengths of the oscillations. Examining the data of the ionosphere, one finds, however, that rather the opposite is true, the mean free path being very often larger than the wavelength divided by 2π. Thus, one cannot use one of the simple pictures just mentioned, and as always when collisions are rare one must turn to a more sophisticated description provided by a kinetic theory. Such a theory consists in Boltzmann-like equations for each particle species, i.e. we describe each species by a density in 6-dimensional phase space $\mathbf{x}$, $\mathbf{v}$. We call it $f_v(\mathbf{x}, \mathbf{v}, t)$, v indicating the species. Corresponding to f_v we have current densities in $\mathbf{x}$- and $\mathbf{v}$-space given by

$$\dot{\mathbf{x}} f_v = \mathbf{v} f_v \quad \text{and} \quad \dot{\mathbf{v}} f_v = (1/m_v)\,\hat{\mathbf{K}}_v f_v,$$

where m_v is the mass of a particle of species v and $\hat{\mathbf{K}}_v$ the force acting on such a particle. A continuity equation in phase space must then hold

$$\frac{\partial f_v}{\partial t} + \frac{\partial}{\partial \mathbf{x}} \cdot (\mathbf{v} f_v) + \frac{\partial}{\partial \mathbf{v}} \cdot \left(\frac{1}{m_v} \hat{\mathbf{K}}_v f_v \right) = 0 .$$

It is the essential approximation of a Boltzmann-like equation that it splits the total force $\hat{\mathbf{K}}_v$ into an average force $\mathbf{K}_v$ and a part which can be attributed to collisions and can then be approximated by some kind of collision term. We have then

$$\frac{\partial f_v}{\partial t} + \mathbf{v} \cdot \frac{\partial f_v}{\partial \mathbf{x}} + \frac{1}{m_v} \mathbf{K}_v \cdot \frac{\partial f_v}{\partial \mathbf{v}} = \left(\frac{\partial f_v}{\partial t} \right)_{\text{coll.}}$$

Here it is assumed that $(\partial/\partial \mathbf{v}) \cdot \mathbf{K}_v$ vanishes, which is true if $\mathbf{K}_v$ is the Lorentz force

$$\mathbf{K}_v = e_v \left(\mathbf{E} + \frac{1}{c} \mathbf{v} \times \mathbf{B} \right).$$

According to the definition of $\mathbf{K}_v$ the electric and magnetic fields $\mathbf{E}$ and $\mathbf{B}$ are average fields; they arise not from single particles but from average particle densities and current densities, where these averages are calculated by the use of f_ν in the form

$$n_\nu(\mathbf{x}, t) = \int f_\nu(\mathbf{x}, \mathbf{v}, t)\, d^3v$$

$$n_\nu \mathbf{v}_\nu(\mathbf{x}, t) = \int \mathbf{v} f_\nu(\mathbf{x}, \mathbf{v}, t)\, d^3v.$$

Inserting these expressions into the Maxwell equations we obtain a closed set of equations for f_ν, **E**, **B**. A theory of this type is called a Vlasov theory if collisions are fully neglected. In the following, collisions of ions with neutrals are taken into account, but collisions between charged particles are neglected because of their low density, and collisions of electrons with neutrals because of the small cross-section. Because of the collisions between ions and neutrals, the ion distribution function, if disturbed, say in the form of a wave, tends to become equal to the distribution of the neutrals. The latter distribution can always be assumed to be a Maxwellian. Thus also the unperturbed ion distribution is a Maxwellian with a temperature T_i. We will first consider waves under the assumption that the unperturbed state is homogeneous in space with an ion density n_0. We then write for the unperturbed ion distribution

$$f_i^0 = n_0 f_i^M(\mathbf{v})$$

with

$$f_i^M(\mathbf{v}) = (m_i/2\pi K T_i)^{3/2} \exp - (m_i v^2 / 2 K T_i)$$

and approximate the collision term for the ions simply by

$$(\partial f_i / \partial t)_{\text{coll}} = - \nu_i (f_i - f_i^0),$$

whereas we choose

$$(\partial f_e / \partial t)_{\text{coll}} = 0$$

or formally

$$(\partial f_e / \partial t)_{\text{coll}} = - \nu_e (f_e - f_e^0),\ \nu_e = 0.$$

We will now investigate the following situation: The unperturbed state is given by f_i^0 for the ions. The electrons are again described by a Maxwellian

$$f_e^0 = n_0 (m_e / 2\pi K T_e)^{3/2} \exp - (m_e / 2 K T_e)(\mathbf{v} - \boldsymbol{\mu})^2$$

with density n_0 and temperature T_e which is generally different from T_i. But we allow this Maxwellian to be shifted by $\boldsymbol{\mu}$ in velocity space in the direction of the ionospheric magnetic field which is assumed to be constant in space and time. Thus we account for currents parallel to the magnetic field, which have been actually observed. We want to find out what kind of waves are possible in such a system, or more specifically what waves will have growing amplitudes, i.e. will be unstable. The latter is important, because perturbations in the form of waves are always present, but with very small amplitudes. Only growing waves will reach amplitudes that can be observed. Of course, such an instability must be due to the current.

I will not give a pure deductive theory here. This would not be very reasonable because of the tremendous number of degrees of freedom contained in our equations. At present one knows of about 50 different instabilities in plasma physics, and probably an infinite number exists. What will be done is to discuss from the beginning those waves which seem the most likely to explain certain phenomena. These are the so-called ion-acoustic waves in which only an electric field perturbation is present, the magnetic field remaining nearly unchanged.

I will try, with an analytical treatment, to give you a feeling about the nature of these waves. The procedure will be first a more formal mathematical one, and then I will show you the physics behind it. What we have to do is to linearize our equations, mainly the Boltzmann equation yielding (with $\mathbf{E}^1 \parallel \mathbf{B}^0$)

$$\frac{\partial f_v^1}{\partial t} + \mathbf{v}\frac{\partial f_v^1}{\partial \mathbf{x}} + \frac{e_v}{m_v}\mathbf{E}^1\frac{\partial f_v^0}{\partial \mathbf{v}} = -v_v f_v^1 \,.$$

In order to find out how an initial perturbation will develop, we perform a Laplace transformation according to

$$f_v^1(\omega), \mathbf{E}^1(\omega) = \int_0^\infty e^{i\omega t} f_v^1(t), \mathbf{E}^1(t)\,\mathrm{d}t, \operatorname{Im}\omega > 0 \text{ sufficiently}$$

$$f_v^1(t), \mathbf{E}(t) = \frac{1}{2\pi}\int_{-\infty+i\gamma}^{+\infty-i\gamma} e^{-i\omega t} f_v^1(\omega), \mathbf{E}^1(\omega)\,\mathrm{d}\omega, \gamma > 0 \text{ sufficiently}\,.$$

Then

$$\int_0^\infty \frac{\partial f_v^1}{\partial t} e^{i\omega t}\,\mathrm{d}t = -f_v^1(t=0) - i\omega f_v^1(\omega)$$

and our linearized equation reads, with

$$f_v^1, \mathbf{E}^1 \sim e^{ikx}; x, v, E^1 \text{ components parallel to the magnetic field } \mathbf{B}^0,$$

$$-i(\omega + iv_v - kv)f_v^1(\omega) + \frac{e_v}{m_v}E^1\frac{\partial f_v^0}{\partial v} = f_v^1(t=0)$$

from which we obtain

$$f_v^1(\omega) = \frac{e_v n_0}{m_v}E^1\frac{1}{i(\omega + iv_v - kv)}\frac{\partial f_v^M}{\partial v} - \frac{f_v^1(t=0)}{i(\omega + iv_v - kv)}\,.$$

The first-order charge-density times 4π follows from this expression as

$$4\pi\varrho^1 = \sum_v 4\pi e_v \int f_v^1\,\mathrm{d}^3 v$$

$$= \sum_v \left[\frac{4\pi e_v^2 n_0}{m_v}\int\frac{\partial f_v^M/\partial v}{i(\omega + iv_v - kv)}\,\mathrm{d}^3 v E^1 - 4\pi e_v\int\frac{f_v^1(t=0)}{i(\omega + iv_v - kv)}\,\mathrm{d}^3 v\right].$$

This is to be inserted into the Poisson equation $ikE^1 = 4\pi\varrho^1$. Free waves are given by solutions of the homogeneous part of this equation yielding the dispersion relation

$$k^2 = \sum_\nu \omega_{p\nu}^2 \int \frac{\partial f_\nu^M/\partial v}{v - \dfrac{\omega + iv_\nu}{k}} \, d^3v, \quad \omega_{p\nu} = \sqrt{(4\pi\, e_\nu^2 n_0/m)}.$$

The integrations over the velocity components perpendicular to the magnetic field can be carried through immediately. If we write

$$f_\nu^M = \left(\frac{m_\nu}{2\pi K T_\nu}\right)^{1/2} \exp -\left[\frac{m_\nu}{2KT}(v-\mu_\nu)^2\right]$$

$$\mu_\nu = \mu \text{ for electrons}$$
$$\quad = 0 \text{ for ions}$$

we find also

$$k^2 = \sum_\nu \omega_{p\nu}^2 \oint_{-\infty}^{\infty} \frac{\partial f_\nu^M/\partial v}{v + \mu_\nu - \dfrac{\omega + iv_\nu}{k}} \, dv,$$

where $k > 0$ has been assumed. In this case the v-integration is to be performed in such a way that, according to the original demand of $\text{Im}\,\omega > 0$, the analytic continuation for $\text{Im}\,\omega < 0$ is obtained; i.e., the path of integration is below $(\omega + iv_\nu)/k$ which is indicated by the hook on the integral sign.
We have

$$\frac{\partial f_\nu^M}{\partial v} = -\frac{m_\nu v}{KT_\nu} f_\nu^M.$$

Introducing

$$\sqrt{\frac{KT_e}{4\pi n_0 e_\nu^2}} = D = \text{electron Debye length}$$

we also find

$$k^2 D^2 = -\sum_\nu \frac{T_e}{T_\nu} \oint_{-\infty}^{\infty} \frac{v}{v + \mu_\nu - \dfrac{\omega + iv_\nu}{k}} f_\nu^M \, dv$$

or with

$$x = \frac{v}{u_\nu}, \quad f^M(x) = \frac{1}{(2\pi)^{1/2}} e^{-x^2/2}$$

$$u_\nu = (KT_\nu/m_\nu)^{1/2} = \text{thermal velocity of species } \nu$$

$$k^2 D^2 = -\sum_\nu \frac{T_e}{T_\nu} \frac{1}{(2\pi)^{1/2}} \oint_{-\infty}^{\infty} \frac{xe^{-x^2/2}}{x - \left(\dfrac{\omega + iv_\nu}{k} - v_\nu\right)\Big/u_\nu} \, dx.$$

Let us now discuss this dispersion relation. First, what do we expect? Let us make the picture as given in Figure 1.

If we had a plasma like this, i.e. like a crystal, we would have two very different branches of oscillations:

(1) Ions are at rest and electrons move; these are the so-called plasma oscillations or Langmuir oscillations. They have a frequency given by ω_{pe}.

(2) Electrons are at rest and ions move; these oscillations should correspondingly have ω_{pi} as frequency.

Because of the large mass of ions the Langmuir oscillation with frequency ω_{pe} can

Fig. 1. Plasma as a crystal.

have any wavelength whereas the ion oscillations with frequencies ω_{pi} are possible only for very short wavelengths; for larger ones these waves are modified and, as we shall see, become soundlike, their frequencies always being small compared with the electron plasma frequency. The electron plasma frequency is much higher than the frequencies observed in the ionosphere. That is the reason why we are interested in ion waves. Observations show in addition that the inequalities

$$u_i < \omega/k \ll u_e$$

hold.

If for the moment we drop μ_y, we can approximately solve our dispersion relation. In the electron term we neglect $(\omega + iv_e)/ku_e$. In the ion term we expand

$$\frac{x}{x - (\omega + iv_i)/ku_i} \simeq - \frac{x}{(\omega + iv_i)/ku_i} - \frac{x^2}{(\omega + iv_i)^2/k^2u_i^2} - \cdots .$$

The dispersion relation therefore reads approximately

$$k^2 D^2 = -1 + \frac{T_e}{T_i} \frac{k^2 u_i^2}{(\omega + iv_i)^2} \frac{1}{(2\pi)^{1/2}} \int\limits_{-\infty}^{\infty} x^2 \exp - (x^2/2)\, dx$$

$$= -1 + \frac{k^2}{(\omega + iv_i)} \frac{KT_e}{m_i}$$

from which

$$(\omega + iv_i)^2 = \frac{k^2 \dfrac{KT_e}{m_i}}{1 + k^2 D^2}$$

follows.

For $kD \gg 1$ we have

$$(\omega + iv_i)^2 \simeq \frac{KT_e/m_i}{KT_e/4\pi n_0 e^2} = \frac{4\pi n_e e^2}{m_i} = \omega_{pi}^2$$

in agreement with our physical discussion. For $kD < 1$ we have, however,

$$\omega + iv_i \simeq k \sqrt{(KT_e/m_i)}.$$

This is a soundlike wave with the pressure given by the electrons and the inertia by the ions.

Whereas in normal sound collisions provide the coupling between particles, here a small collective electric field couples the electrons to the ions, the energy in this electric field being negligible compared with the energy in the oscillation of electrons and ions, as exhibited by the dispersion relation. I shall come back to this point, which is very characteristic for soundlike waves.

What we have done so far was not quite correct. The result could have been obtained equally well with the two-fluid theory. In using the expansion for the ions, we have

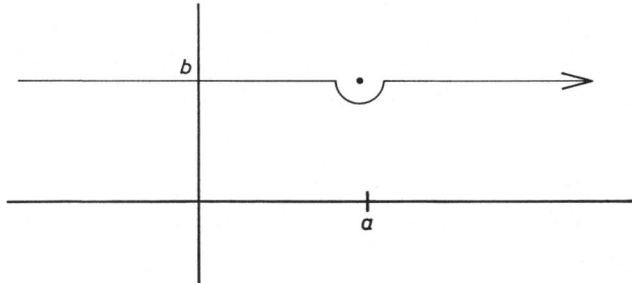

Fig. 2.   Path of integration.

neglected the main kinetic effect which consists in the resonance shown up by the zero of the denominator in the integral. This zero obviously means a resonance between the motion of certain particles and the wave propagation. Let us therefore investigate this part of the integral in more detail.

The form of our integral is

$$\oint_{-\infty}^{\infty} \frac{g(x)}{x - a - ib} \, dx = \oint_{-\infty}^{\infty} \frac{g(y + a + ib)}{y} \, dy.$$

The resonance part contributes because we must not only take the principal value of the integral but must in addition go around the singularity, say on a half circle with infinitesimally small radius (Figure 2).

This half circle gives

$$i\pi g(a + ib).$$

The full dispersion relation therefore reads (P: principal value)

$$
k^2 D^2 = -\sum_v \frac{T_e}{T_v} \frac{1}{(2\pi)^{1/2}} \left[P \int_{-\infty}^{\infty} \frac{x e^{-x^2/2}}{x - \left(\frac{\omega + i v_v}{k} - \mu_v \right) / u_v} \, dx \right.
$$
$$
\left. + i\pi \frac{1}{(2\pi)^{1/2}} \frac{\frac{\omega + i v_v}{k} - \mu_v}{u_v} \cdot \exp \left\{ - \frac{\left(\frac{\omega + i v_v}{k} - \mu_v \right)^2}{2 u_v^2} \right\} \right].
$$

If we again take $\mu_v = 0$ and do the same approximations as before for the principal part of the integral only, which, however, is now really justified, we get

$$
k^2 D^2 = -1 - i\pi \frac{1}{(2\pi)^{1/2}} \frac{\omega}{k u_e} \exp - (\omega^2/2k^2 u_e^2) + \frac{k^2}{(\omega + i v_i)^2} \frac{K T_e}{m_i}
$$
$$
- i\pi \frac{1}{(2\pi)^{1/2}} \frac{\omega + i v_i}{k u_i} \exp - ((\omega + i v_i)^2/2k^2 u_i^2).
$$

The additional terms are indeed small, but they are important because they contribute to the imaginary part of ω which determines whether damping or growth of the waves occurs. Inserting in these expressions the former expression for ω as a first approximation, we obtain with $v_i = 0$

$$
\omega \simeq k \left(\frac{K T_e}{m_i} \right)^{1/2} - i \frac{1}{2} \left(\frac{\pi}{2} \right)^{1/2} \frac{(m_e/m_i)^{1/2}}{(1 + k^2 D^2)^2} k \left(\frac{K T_e}{m_i} \right)^{1/2}
$$
$$
\exp - ((m_e/2m_i)(1/(1 + k^2 D^2)) - i \frac{1}{2} \left(\frac{\pi}{2} \right)^{1/2} \frac{(T_e/T_i)^{1/2}}{(1 + k^2 D^2)^2}
$$
$$
k \left(\frac{K T_e}{m_i} \right)^{1/2} \exp - ((T_e/2T_i)(1/1 + k^2 D^2)).
$$

Thus the wave is now damped without having collisions, the damping usually being due mainly to the resonance of ions with the wave. Such collisionless damping is called Landau damping.

We now of course expect that by introducing μ, damping may be changed into growth. Again taking $v_v = 0$, the only imaginary part of the dispersion relation for real ω then stems from the new terms, which must therefore be cancelled if we want to find the transition point from damping to growth, i.e.

$$
\frac{1}{u_e} \left(\frac{\omega}{k} - \mu_e \right) \exp - ((\omega/k - \mu_e)^2/2u_e^2) = -\frac{1}{u_i} \left(\frac{\omega}{k} \right) \exp - (\omega^2/(2k^2 u_i^2))
$$

or

$$
\mu_e = \frac{\omega}{k} + \frac{u_e}{u_i} \frac{\omega}{k} \exp - (\tfrac{1}{2}(\omega^2/k^2 u_i^2) - (\omega - k\mu_e)^2/k^2 u_e^2)).
$$

We expect $|\omega/k - \mu_e| \ll \mu_e$ in which case

$$\mu_e \simeq \frac{\omega}{k} + \frac{u_e}{u_i}\frac{\omega}{k} \exp - \left(\tfrac{1}{2}(\omega^2/k^2 u_i^2)\right).$$

Inserting $\omega/k = (KT_e/m_i)^{\frac{1}{2}}(1/(1 + k^2 D^2)^{\frac{1}{2}})$ we arrive at

$$\mu_e = \left(\frac{KT_e}{m_i}\right)^{1/2} \frac{1}{(1 + k^2 D^2)^{1/2}}$$

$$\left\{1 + \left(\frac{m_i T_e}{m_e T_i}\right)^{1/2} \exp - \left(\tfrac{1}{2}(T_e/T_i)/(1 + k^2 D^2)\right)\right\}.$$

Owing to our expansion we have $1 + k^2 D^2 \ll T_e/T_i$. Thus for $T_e/T_i = 3$, for instance, $k^2 D^2$ should be smaller than 1. We then have approximately

$$\mu_e \simeq (KT_e/m_e)^{1/2}(T_e/T_i)^{1/2} \exp - \left(\tfrac{1}{2}(T_e/T_i)/(1 + k^2 D^2)\right).$$

The minimum value is reached for $k = 0$ and is

$$\mu_{e\mathrm{Min}} \simeq 0.4(KT_e/m_e)^{1/2}.$$

The integrals involved here are tabulated by Conte and Fried. With these tables you obtain the more correct value

$$\mu_{e\mathrm{Min}} = 0.3(KT_e/m_e)^{1/2}.$$

In the present case μ_e had to be only large enough to overcome Landau damping. If we take into account ion collisions, μ_e must be larger. D'Angelo has done the calculations for different values of $\varepsilon = D/\lambda_{\mathrm{in}}$, where λ_{in} is the mean free path for ion-neutral collisions. He will report on his results and their consequences.

Until now we have discussed only currents parallel to the magnetic field. There are of course also currents perpendicular to the field, predominantly at higher altitudes where we have a plasma being confined by the earth magnetic field. The situation is roughly the one as given in Figure 3.

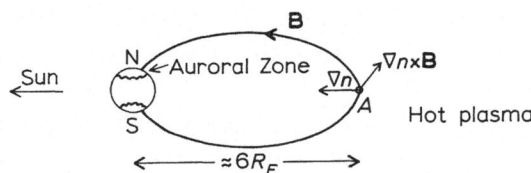

Fig. 3.   Earth's magnetic field configuration up to an altitude of
about 6 earth radii.

In the neighbourhood of point A we choose: x-direction parallel to ∇n, y-direction parallel to $\nabla \mathbf{n} \times \mathbf{B}$, z-direction parallel to $\mathbf{B}$.

Let us analyze in a very simplified way the effect of these currents on waves.

We are again interested in waves with $|\omega/k_\parallel| \ll u_e$, ($\parallel$: parallel to $\mathbf{B}^0$). Neglecting again electron collisions and $\partial/\partial t = -i\omega$ compared with $\mathbf{v} \cdot (\partial/\partial \mathbf{x}) = 0$ ($k_\parallel u_e$), our nonlinearized Vlasov equation for the electrons reads approximately

$$\mathbf{v} \cdot (\partial f_e / \partial x) + \frac{1}{m_e} K_e (\partial f_e / \partial v) = 0$$

of which

$$f_e = n^0 \left(\frac{m_e}{2\pi K T_e} \right)^{3/2} \exp - \left((1/K T_e) \left(\tfrac{1}{2} m_e v^2 - e\Phi \right) \right)$$

is a solution yielding

$$n_e^1 / n^0 = e\Phi^1 / K T_e .$$

(Φ is the electric potential corresponding to K_e.) This is exactly the same result which we would have obtained using the approximation in the former procedure. It describes the typical behaviour of the electrons in slow waves, i.e. ion waves, where one can assume the electrons to be always in thermodynamic equilibrium. There is an interesting consequence of this relation. With

$$k^2 \Phi^1 = 4\pi \, e (n_i^1 - n_e^1) = (k^2 K T_e / en^0) \, n_e^1$$

the relation

$$(k^2 K T_e / 4\pi n_0 \, e^2) = k^2 D^2 = (n_i^1 - n_e^1)/n_e^1$$

holds.

Soundlike waves are characterized by $k^2 D^2 \ll 1$; these waves are therefore quasi neutral'. They can be approximated by the condition $n_i^1 = n_e^1$ instead of using Poisson's equation. This is in agreement with the fact, which we have already observed, that the electrical energy in such waves is negligible.

Let us go on and look at the ions. We have learned in the above treatment that, for $|\omega/k| \gg u_i$, thermal effects for the ions can be neglected entirely as long as we do not discuss damping or growth. The ions being at rest in the unperturbed state, the linearized equation of motion in the perturbed state follows

$$(\partial \mathbf{v}_i^1 / \partial t) = - \frac{e}{m_i} \nabla \Phi^1 + (e/m_i c) \, \mathbf{v}_i^1 \times \mathbf{B}^0$$

or in WKB approximation, assuming only small gradients of the unperturbed quantities,

$$- i\omega \mathbf{v}_i^1 = - i (e/m_i) \, \mathbf{k} \Phi^1 + (e/m_i c) \, \mathbf{v}_i^1 \times \mathbf{B}^0 .$$

From this we find

$$v_{i\parallel}^1 = (ek_\parallel / m_i \omega) \, \Phi^1$$

and

$$\mathbf{v}_{i\perp}^1 = \left(\mathbf{k}_\perp + i \frac{\omega_c}{\omega} \mathbf{k}_\perp \times \mathbf{b}^0 \right) \frac{1}{1 - \dfrac{\omega_c^2}{\omega^2}} \frac{e}{m_i \omega} \Phi^1$$

$$\simeq - (e/m_i \omega_c) \, \mathbf{k}_\perp \times \mathbf{b}^0 \Phi^1 ; \qquad \omega_c = eB_0 / m_i c ,$$

where $\mathbf{b}^0$ is the unit vector in the direction of $\mathbf{B}^0$ and $\omega \ll \omega_c$. We can insert these expressions into the continuity equation for the ions

$$\partial n_i^1/\partial t + \nabla \cdot (n_i^0 \mathbf{v}_i^1) = 0 .$$

Again using the WKB approximation yields, inserting the expression for $\mathbf{v}_i^1$,

$$n_i^1 = \frac{n_e}{m_i \omega^2} \left\{ k_\parallel^2 - k_y K_x \frac{\omega}{\omega_c} \right\} \Phi^1 , \quad K_x = \frac{1}{n_i^0} \frac{dn_i^0}{dx} .$$

For $kD_e < 1$ we can use the quasi-neutral approximation $n_i^1 = n_e^1$ instead of Poisson's equation as shown above. This leads to the dispersion relation

$$\omega \simeq -\tfrac{1}{2}(v_s^2 k_y K_x/\omega_c) \pm \sqrt{(\tfrac{1}{4}(v_s^2 k_y K_x/\omega_c)^2 + v_s^2 k_\parallel^2)}, \quad v_s^2 = (KT_e/m_i) .$$

For $k_\parallel = 0$ we obtain $\omega = 0$ and $\omega = -(v_s^2 K_x/\omega_c)k_y \equiv \omega^*$.

The frequency ω^* is a very interesting one. Let us consider the equilibrium equation (remember that we neglected the ion temperature).

$$(1/c)\,\mathbf{j}^0 \times \mathbf{B}^0 = \nabla p^0 = KT_e \nabla n_e^0 = n^0 KT_e K_x .$$

We have $(1/c)(\mathbf{j}^0 \times \mathbf{B}^0)_x = (1/c)j_y^0 B^0 = -nv_D e$ where v_D is the drift velocity of the electrons. Therefore

$$v_D = -\left(KT_e/(eB^0/c)\right) K_x = -v_s^2 K_x/\omega_c .$$

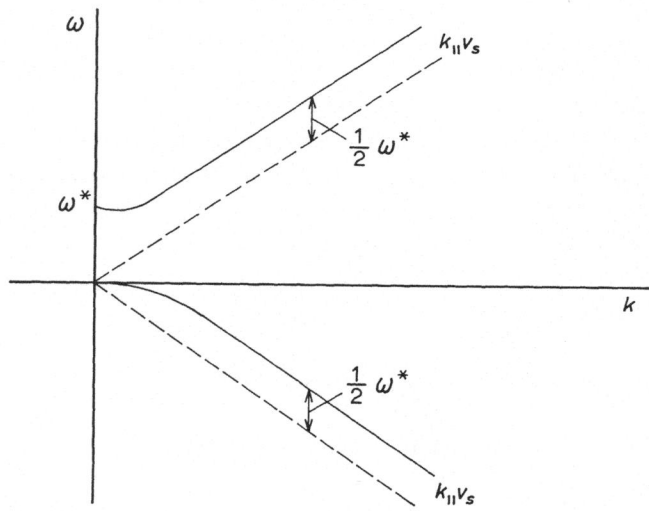

Fig. 4. Dispersion relation for drift waves.

Thus our frequency ω^* can also be written as $\omega^* = k_y v_D$, i.e. the wave pattern moves along with the drifting electrons. For this reason one calls such waves drift waves. v_D is usually a rather small velocity, much smaller than v_s. Therefore both wave types with $\omega = 0$ and $\omega = \omega^*$ for $k_\parallel = 0$ very soon become ion sound waves for $k_\parallel \neq 0$ with $\omega = k_\parallel v_s$ (Figure 4).

For this reason the first one is also called decelerated ion sound wave and the second accelerated ion sound wave where the change in the velocities is due to the electron drift.

We have so far made use of a very rough picture. We may, however, expect a Vlasov treatment of the phenomena to show up instabilities for the resonant waves, i.e. for accelerated ion sound waves. I will not go through the details of such calculations which are rather lengthy but straightforward. Under the conditions

$$\beta = n^0 KT/(B_0^2/8\pi) < m_e/m_i, \quad u_e > |\omega/k_\parallel| > u_i, \quad k_\parallel^2 \ll k_y^2$$
$$T_i = T_e$$

one arrives at the dispersion relation

$$\omega = k_y v_D \frac{e^{-Z}I_0(Z)}{2 - e^{-Z}I_0(Z)} + i \frac{2}{\sqrt{\pi}} \frac{(k_y v_D)^2}{|k_\parallel| \sqrt{(2KT/m_e)}} \frac{e^{-Z}I_0(Z)(1 - e^{-Z}I_0(Z))}{[2 - e^{-Z}I_0(Z)]^3}$$

with $Z = k_y^2 r_i^2$ and $r_i^2 = $ square of the thermal ion gyroradius $= m_i KTc^2/e^2 B_0^2$. I_0 is the Bessel function of imaginary argument of order zero.

This relation will be used by D'Angelo. It is restricted in the sense of the inequalities. A main point is the smallness of β which excludes coupling of Alfvén waves to our pure electrostatic waves.

GENERAL FEATURES AND SATELLITE OBSERVATIONS OF MAGNETOIONIC AND MAGNETOHYDRODYNAMIC WAVES IN THE OUTER IONOSPHERE

H. KIKUCHI

Max-Planck-Institut für Aeronomie, Institut für Ionosphären-Physik, Lindau/Harz, Germany

Abstract. General features of magnetoionic and magnetohydrodynamic waves including plasma waves in the outer ionosphere are reviewed on the basis of a combined electromagnetic and plasma kinetic theory with particular reference to satellite observations. Explicit expressions for the refractive indices of these waves in the outer ionosphere are presented taking into account the temperature effects and Landau and cyclotron damping for both electrons and ions in a generalized form of the Appleton–Hartree equations. Formulas are specifically provided for the wave features in the HF, VLF, and ELF ranges in conjunction with topside soundings, whistlers, micropulsations, and VLF and ELF radiations. Further, possible types of plasma resonance in the ionosphere are considered over a whole range of frequencies, with some experimental data observed mainly by Alouette and partly by Injun. In this connection, the satellite experiments on micropulsations, plasma resonances and radiations in the ELF range should be included as a method supplementing ground-based soundings.

1. Introduction

The use of space rockets and satellites to investigate the structure and properties of the top-side ionosphere and interplanetary space has brought a number of new observations of magnetoionic, magnetohydrodynamic and plasma waves in the outer ionosphere over a wide range of frequencies.

The artificial swept radio-frequency soundings carried out with the top-side sounder Alouette have resulted in new observations of magnetoionic and plasma waves in the HF range, for instance a new type of slow Z-waves and plasma, cyclotron, and upper hybrid resonances (Warren, 1962, 1963; Lockwood, 1963), while passive Alouette, Injun and some other satellites have received a new class of magnetoionic, magneto-hydrodynamic and plasma waves in the VLF and ELF ranges, for instance the so-called non-ducted whistlers, electron and ion whistlers, and lower hybrid resonances (Smith *et al.*, 1964; Gurnett *et al.*, 1965; Brice and Smith, 1965; Barrington *et al.*, 1965).

Furthermore, it has recently been recognized that many basic geophysical phenomena such as geomagnetic pulsations, disturbances, and solar wind–magnetosphere interactions are associated more or less with magnetohydrodynamic waves or the so-called 'Alfvén' waves, and that geomagnetic pulsations, for instance, are manifestations of Alfvén waves in the magnetosphere (Troitskaya and Gul'elmi, 1967). These findings have not always been the outcome of satellite experiments.

In order to follow up these new experimental results, neither the normal magnetoionic theory based on the Appleton–Hartree equations in a cold plasma nor the conventional magnetohydrodynamic theory in a conducting fluid is suitable except for very limited conditions. Furthermore, theory of pure plasma waves normally

D'Angelo (ed.), Low-Frequency Waves and Irregularities in the Ionosphere. All rights reserved

neglects their coupling with magnetoionic and/or magnetohydrodynamic waves which becomes significant in the vicinity of resonances in a magnetoplasma or in some magnetosonic regime where the Alfvén speed is small compared to the electron or ion thermal velocity.

A more general theory to meet these requirements is outlined on the basis of a combined electromagnetic and plasma kinetic theory (Stix, 1962; Kikuchi, 1963, 1964) in a generalized form of the Appleton–Hartree equations familiar to ionospheric workers, and is applied to some analyses and interpretations of several ionograms and spectrograms recorded by 'Alouette', Injun and so on. From this point of view, magnetoionic and magnetohydrodynamic waves as defined in this article include electron and ion plasma waves.

It should be noted that the recent comprehensive review of 'VLF and ELF Waves in the Near Earth Plasma' by Al'pert (1967) is based on a plasma kinetic theory developed largely by Russian scientists.

2. Basic Plasma Parameters and Properties

In this section we give a brief review of a variety of waves in the outer ionosphere over a whole range of frequencies and summarize basic plasma parameters and properties of the outer ionosphere.

Table I shows a classification of possible types of wave. The magnetoionic waves normally include high frequency radio waves in addition to the waves in the VLF range which is sometimes called the 'whistler' range, while the ELF range is that for the magnetohydrodynamic waves or the so-called 'micropulsations'. In contrast with artificial radio soundings of importance in the HF range, natural electromagnetic phenomena in the magnetosphere play a significant role in the VLF and ELF ranges. In this table, some modes, the electron-plasma waves ion-plasma waves, and magnetosonic waves, are new appearances due to the thermal or temperature effects of plasmas. In addition, a hot plasma theory imposes some thermal corrections together with Landau type damping and normal collisional damping upon the electromagnetic and magnetohydrodynamic waves. When treating a plasma, we must bear in mind the relative importance of the thermal and collisional effects. From these and geophysical points of view, we are interested in the altitude dependence of basic plasma parameters.

Figures 1 and 2 show the collision frequencies of various kinds with the plasma and cyclotron frequencies in the height range of 10^2 to 10^6 km and indicate that the collisional effects decrease rapidly with increasing height in the outer ionosphere, although the electron–ion (+neutral) collision frequency and the ion–ion collision frequency possess a maximum of 3 kc/s or 15 c/s respectively at a height of 300 km. The collisional effects predominate when the wave frequency becomes comparable with the collision frequency and in general become more important for the bottom side ionosphere than for the outer ionosphere. The electron plasma frequency normally exceeds the electron cyclotron frequency but the former can become lower than the

TABLE I

Classification of waves in the magnetosphere

Frequency range		Type of wave	Polarization	Phase velocity	Damping
HF	$\omega \gtrsim \omega_{p-}, \omega_1$	Ordinary	L-mode	$v > c_0$	Electron cyclotron
		Extraordinary	R-mode (L-mode)	$v > c_0$; $c_0 \gtrsim v \gg v_-$	Electron Landau; Electron cyclotron
Magneto-ionic waves (Whistlers)	VLF $\omega_{c-} \gtrsim \omega > \sqrt{\omega_{c+}\omega_{c-}}$	*Electron plasma*		$v \gtrsim v_-$	Electron cyclotron
		Extraordinary (modified Alfvén, fast Alfvén)	R-mode	$v_A/2\sqrt{\eta} \geq v \geq v_A\,(0 \leq \theta < \pi/2)$; $v_A \geq v > v_s \gtrsim v_+\,(\theta \lesssim \pi/2)$	Ion and electron Landau
		Fast ion plasma		$v_- > v \gtrsim v_s \gtrsim v_+$	Ion cyclotron Ion and electron Landau
Magneto-hydro-dynamic waves	ELF $0 \lesssim \omega < \sqrt{\omega_{c+}\omega_{c-}}$	Ordinary (pure Alfvén, slow Alfvén)	L-mode	$v \lesssim v_A$	Ion and electron Landau
		Extraordinary (modified Alfvén, fast Alfvén, fast magnetosonic)	R-mode	$v \gg v_A$	Ion and electron Landau
(Micropulsations)		*Slow ion plasma* (slow magnetosonic)		$v_A > v \gtrsim v_s \gtrsim v_+$	Ion and electron Landau

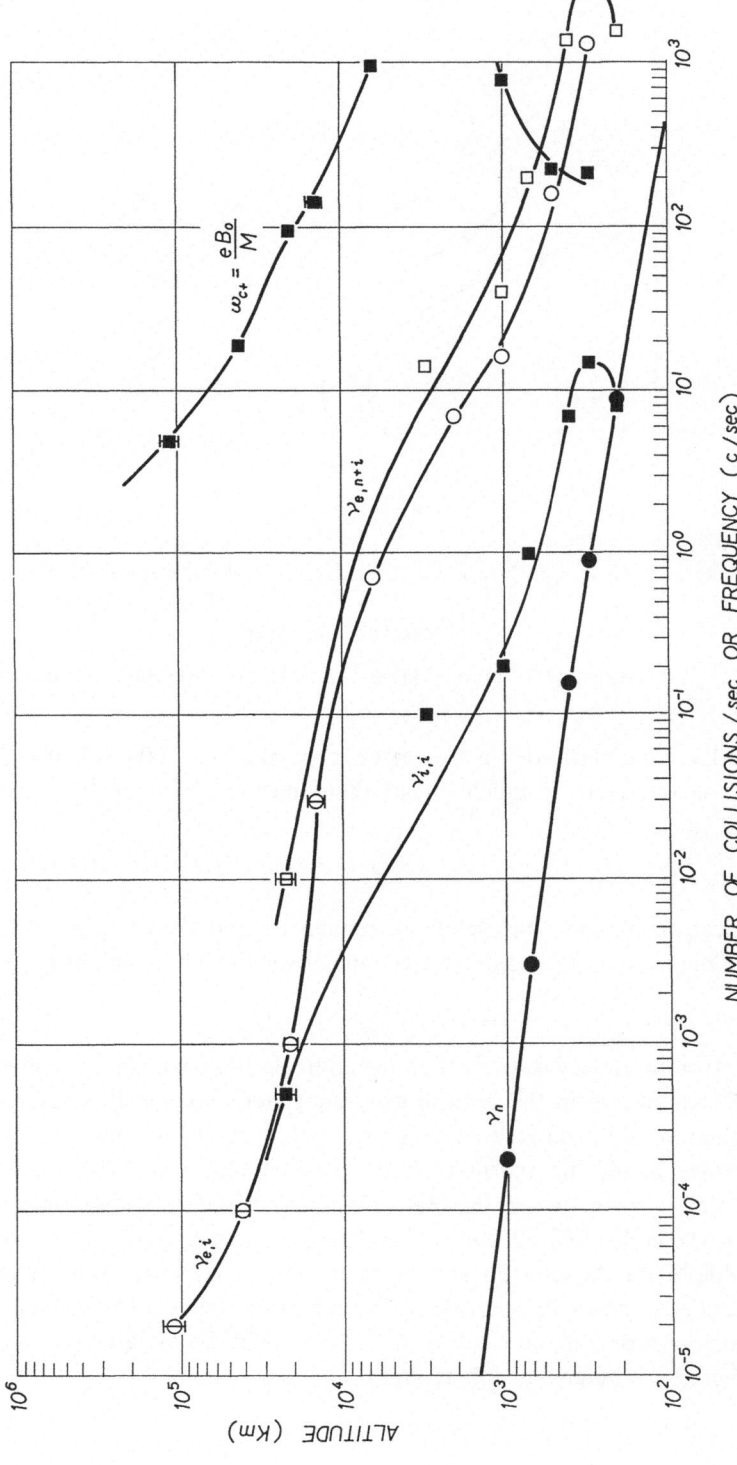

Fig. 1. Altitude profiles of number of collision, plasma, and cyclotron frequencies.

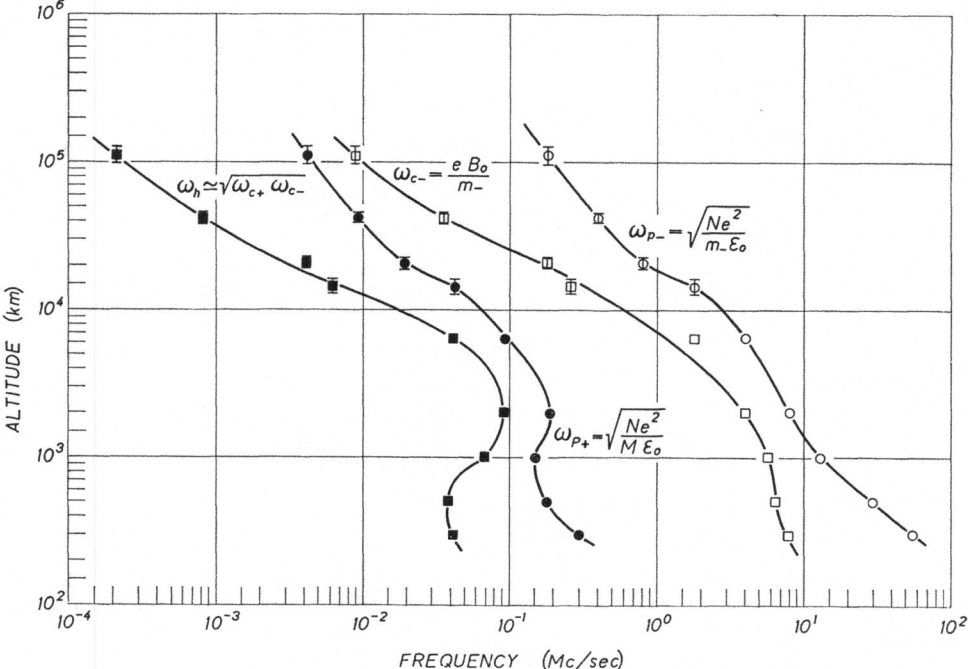

Fig. 2.   The plasma, cyclotron, and lower hybrid frequency profiles (continued).

latter at middle or high latitudes in the height range of about 1 000 to 10 000 km, while the ion plasma frequency is much higher than the ion cyclotron frequency in the whole height range, i.e.

$$\omega_{p+} \gg \omega_{c+}.\tag{1}$$

While the electron plasma and cyclotron frequencies and the ion plasma frequency are all much higher than the collision frequency over the whole height range, i.e.

$$\omega_{p-}, \, \omega_{c-}, \, \omega_{p+} \gg \nu_c \simeq \nu_{ei},\tag{2}$$

the ion cyclotron frequency is lower than the collision frequency below 600 ~ 800 km, i.e. in the F_2-region and in the bottom side ionosphere, so that it is rather difficult to observe the ion cyclotron resonances in and below the F_2-region.

On the other hand, the thermal effects predominate when the wave velocity approaches the electron or ion thermal velocity and in general become more significant than the collisional effects as the altitude is increased. Figure 3 shows the altitude profile of the Alfvén, ion and electron thermal velocities and indicates that the thermal effects gradually increase with increasing height. The Alfvén speed is beyond or of the order of the electron thermal velocity below a height of 20 000 ~ 30 000 km, i.e. in the outer ionosphere and on its boundary

$$v_A \gtrsim v_{T-},\tag{3}$$

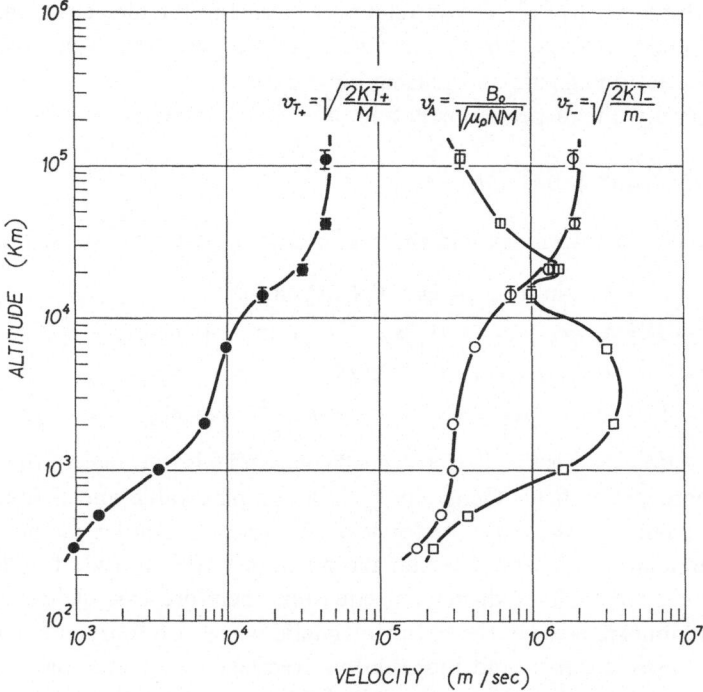

Fig. 3. Alfvén and thermal velocity profiles with altitude.

where

$$v_A = \frac{c}{n_A} = \frac{B_0}{\sqrt{\mu_0(N_+ m_+ + N_- m_-)}} \simeq c \frac{\omega_{c+}}{\omega_{p+}}, \tag{4}$$

c is the velocity of light, n_A the Alfvén refractive index, B_0 the earth's magnetic field, N and m number and mass of particles. The subscripts $+$, $-$ refer to ions and electrons, respectively. Rationalized MKS units are used throughout. The Alfvén speed has a maximum at a height of 3000 to 4000 km. This indicates the possibility of a hydromagnetic waveguide between this height and a base of the F_2-region.

Above a height of 20000 to 30000 km, i.e. in the region of transition from the magnetosphere to the interplanetary space, the Alfvén velocity falls rapidly below the electron thermal velocity and even approaches the ion thermal velocity, i.e.

$$v_A \ll v_{T-}. \tag{5}$$

This implies that modified Alfvén waves couple to fast ion plasma waves and indicates the possibility of excitation of the so-called fast and slow 'magnetosonic' waves, besides the pure Alfvén waves, in this region.

3. Dispersion Relations and Refractive Indices

In order to investigate a wide variety of waves in the outer ionosphere, it is necessary to begin with the general dispersion relation in terms of the dielectric tensor relevant for an ionospheric plasma. For this purpose, we bring here the results obtained from a combined electromagnetic and plasma kinetic theory.

Most generally, the dispersion equation of an ionospheric plasma can be expressed as

$$T = An^4 + Bn^2 + C = 0,\tag{6}$$

where n is the refractive index and the coefficients A, B, C may be defined as

$$\begin{aligned}
A &= \varepsilon_{11}\sin^2\theta + \varepsilon_{33}\cos^2\theta + \varepsilon_{13}\sin 2\theta,\\
-B &= \varepsilon_{11}\varepsilon_{33} + \varepsilon_{13}^2 + (\varepsilon_{11}\varepsilon_{22} + \varepsilon_{12}^2)\sin^2\theta + (\varepsilon_{22}\varepsilon_{33} + \varepsilon_{23}^2)\cos^2\theta\\
&\quad + (\varepsilon_{22}\varepsilon_{13} - \varepsilon_{12}\varepsilon_{23})\sin 2\theta,\\
C &= |\varepsilon_{ik}| = (\varepsilon_{11}\varepsilon_{22} + \varepsilon_{12}^2)\varepsilon_{33} + \varepsilon_{11}\varepsilon_{23}^2 - \varepsilon_{22}\varepsilon_{13}^2 + 2\varepsilon_{12}\varepsilon_{13}\varepsilon_{23}.
\end{aligned}\tag{7}$$

Here $\varepsilon_{ik} = 1 + i(\sigma_{ik}/\omega\varepsilon_0)$ is the dielectric coefficient and θ is the angle of the wave vector with the geomagnetic field. Generally ε_{ik} is a complex value and a function of the angular frequency ω, the wave vector $\mathbf{k}$ or the refractive index n, the ion and the electron temperature T_+ and T_-, and the position $\mathbf{r}$ (the altitude for the case of a stratified ionosphere). This dispersion equation therefore can include all types of waves in the ionosphere, i.e. the electromagnetic waves, ordinary and extraordinary, the plasma waves, electron and ion, the low frequency magnetoionic and magneto-hydrodynamic waves, the pure and modified Alfvén waves and magnetosonic waves, the gravity waves and so on, if the dielectric or conductivity tensor σ_{ik} is suitably defined. This can be done for instance with the aid of a combined electromagnetic and plasma kinetic theory which follows hereafter.

For regions of the ionosphere where the conditions of a quasi-Maxwellian and a quasi-cold plasma are satisfied, i.e.

$$\begin{aligned}
|\sqrt{\mu_\pm}| &= \left|\frac{kv_{T\pm}}{\omega_{c\pm}}\sin\theta\right| = \left|\frac{\omega}{\omega_{c\pm}}n\beta_\pm\sin\theta\right| = \left|\frac{\omega}{\omega_{c\pm}}n\sqrt{\epsilon_\pm}\sin\theta\right|\\
&= \left|\frac{\omega}{\omega_{c\pm}}\cdot\frac{v_{T\pm}}{v_p}\sin\theta\right| \ll 1,
\end{aligned}\tag{8}$$

$$\begin{aligned}
|z_{p\pm}| &= \left|\frac{\omega \mp p\omega_{c\pm}}{\sqrt{2}\,\lambda_\pm}\right| = \left|\frac{\omega \mp p\omega_{c\pm}}{\sqrt{2}\,\omega n\beta_\pm\cos\theta}\right| = \left|\frac{\omega \mp p\omega_{c\pm}}{\sqrt{2\epsilon_\pm}\,n\omega\cos\theta}\right|\\
&= \left|\frac{\omega \mp p\omega_{c\pm}}{\sqrt{2}\,\omega\cos\theta}\cdot\frac{v_p}{v_\pm}\right| \gg 1, \quad p = 0, \pm 1, \pm 2, \dots,
\end{aligned}\tag{9}$$

the dielectric tensor may be written as

$$\tilde{\varepsilon} = \tilde{\varepsilon}^{(0)} + n^2(\epsilon_+\tilde{\delta}_+ + \epsilon_-\tilde{\delta}_-),\tag{10}$$

where $\tilde{\varepsilon}^{(0)}$, $\tilde{\delta}_+$, and $\tilde{\delta}_-$ represent the cold plasma components and the coefficients of ion and electron thermal corrections, respectively, $\epsilon_\pm = v_{T\pm}^2/c^2$, $v_{T\pm}^2 = KT_\pm/m_\pm$, $\beta_\pm = v_{T\pm}/c$, c the velocity of light, v_T the thermal velocity, v_p the wave phase velocity,

K the Boltzmann constant, m the particle mass, and the subscripts, $+$ and $-$ refer to ions and electrons.

Explicit expressions for the dielectric coefficients of a cold plasma are well known as

$$\varepsilon_{11}^{(0)} = \varepsilon_{22}^{(0)} = 1 - \sum_{+,-} \frac{\omega_{p\pm}^2}{\omega^2 - \omega_{c\pm}^2},$$

$$\varepsilon_{33}^{(0)} = 1 - \sum_{+,-} \frac{\omega_{p\pm}^2}{\omega^2},$$

$$\varepsilon_{12}^{(0)} = -\varepsilon_{21}^{(0)} = i \sum_{+,-} (\mp) \frac{\omega_{p\pm}^2 \omega_{c\pm}/\omega}{\omega^2 - \omega_{c\pm}^2},$$

$$\varepsilon_{13}^{(0)} = \varepsilon_{31}^{(0)} = \varepsilon_{23}^{(0)} = \varepsilon_{32}^{(0)} = 0, \tag{11}$$

where ω_p and ω_c are the local plasma and cyclotron frequencies and are given by

$$\omega_{p\pm} = |Z_\pm| e \sqrt{\frac{N_\pm}{\varepsilon_0 m_\pm}},$$

$$\omega_{c\pm} = \frac{|Z_\pm| e B_0}{m_\pm}, \tag{12}$$

where $e = |e|$ is the electronic charge, Z the charge number (set equal to 1 for electrons), ε_0 the permittivity of free space, and N the particle density. The summation is taken over all species of ions (referred to as $+$) and electrons (referred to as $-$). For the double sign the upper sign refers to ions and the lower to electrons.

Explicit expressions for the ion and electron thermal corrections are obtained from a Fourier transform solution of the Boltzmann equation together with the full set of Maxwell's equations by employing the asymptotic expansion of the complex error function in a general formula of the dielectric tensor, i.e. for a neutral plasma with electrons $(-)$ and a single kind of ions $(+)$, which is normally satisfied for the HF and VLF ranges throughout the outer ionosphere and for the ELF range above a height of 1000 to 1500 km,

$$\delta_{11\pm} = -\frac{\omega_{p\pm}^2 \omega^2}{\omega^2 - \omega_{c\pm}^2} \left\{ \frac{3 \sin^2 \theta}{\omega^2 - 4\omega_{c\pm}^2} + \frac{\omega^2 + 3\omega_{c\pm}^2}{(\omega^2 - \omega_{c\pm}^2)^2} \cos^2 \theta \right\},$$

$$\delta_{22\pm} = -\frac{\omega_{p\pm}^2 \omega^2}{\omega^2 - \omega_{c\pm}^2} \left\{ \frac{\omega^2 + 3\omega_{c\pm}^2}{(\omega^2 - \omega_{c\pm}^2)^2} \cos^2 \theta + \frac{\omega^2 + 8\omega_{c\pm}^2}{\omega^2(\omega^2 - 4\omega_{c\pm}^2)} \sin^2 \theta \right\},$$

$$\delta_{33\pm} = -\omega_{p\pm}^2 \left(\frac{3}{\omega^2} \cos^2 \theta + \frac{\sin^2 \theta}{\omega^2 - \omega_{c\pm}^2} \right),$$

$$\delta_{12\pm} = -\delta_{21\pm} = \mp i \frac{\omega_{p\pm}^2 \omega_{c\pm} \omega}{\omega^2 - \omega_{c\pm}^2} \left\{ \frac{6 \sin^2 \theta}{\omega^2 - 4\omega_{c\pm}^2} + \frac{3\omega^2 + \omega_{c\pm}^2}{(\omega^2 - \omega_{c\pm}^2)^2} \cos^2 \theta \right\},$$

$$\delta_{13\pm} = \delta_{31\pm} = -\frac{\omega_{p\pm}^2 \omega^2}{(\omega^2 - \omega_{c\pm}^2)^2} \sin 2\theta,$$

$$\delta_{23\pm} = -\delta_{32} = \pm\, i\,\frac{\omega_{p\pm}^2 \omega_{c\pm}(3\omega^2 - \omega_{c\pm}^2)}{2\omega(\omega^2 - \omega_{c\pm}^2)^2}\sin 2\theta.\tag{13}$$

Using Equations (10) to (13), the dispersion Equation (6) may be written, under the quasi-cold conditions (8) and (9), as

$$
\begin{aligned}
&(\epsilon_+ A_{1+} + \epsilon_- A_{1-})n^6 + (A_0 + \epsilon_+ B_{1+} + \epsilon_- B_{1-})\,n^4\\
&+ (B_0 + \epsilon_+ C_{1+} + \epsilon_- C_{1-})n^2 + C_0\\
&= (\epsilon_+ A_{1+} + \epsilon_- A_{1-})\cdot(n^2 - n_o^2)(n^2 - n_e^2)(n^2 - n_p^2) = 0,
\end{aligned}\tag{14}
$$

where the refractive indices, n_o, n_e, n_p denote the ordinary, extraordinary, and plasma waves, and the coefficients $A_{1\pm}$, $B_{1\pm}$, $C_{1\pm}$ are the first order thermal corrections defined by

$$
\begin{aligned}
A_{1\pm} &= \delta_{11\pm}\sin^2\theta + \delta_{33\pm}\cos^2\theta + \delta_{13\pm}\sin 2\theta,\\
-B_{1\pm} &= \varepsilon_{11}^{(0)}\delta_{33\pm} + \varepsilon_{33}^{(0)}\delta_{11\pm} + \{\varepsilon_{11}^{(0)}(\delta_{11\pm} + \delta_{22\pm}) + 2\varepsilon_{12}^{(0)}\delta_{12\pm}\}\sin^2\theta\\
&\quad + (\varepsilon_{11}^{(0)}\delta_{33\pm} + \varepsilon_{33}^{(0)}\delta_{22\pm})\cos^2\theta + (\varepsilon_{11}^{(0)}\delta_{13\pm} - \varepsilon_{12}^{(0)}\delta_{23\pm})\sin 2\theta,\\
C_{1\pm} &= \varepsilon_{33}^{(0)}\{\varepsilon_{11}^{(0)}(\delta_{11\pm} + \delta_{22\pm}) + 2\varepsilon_{12}^{(0)}\delta_{12\pm}\}.
\end{aligned}\tag{15}
$$

First, we have the zeroth order approximation for a cold plasma in the form

$$
\begin{aligned}
n_{o,e}^{(0)2} &= \frac{1}{2A_0}\left(-B_0 \pm \sqrt{B_0^2 - 4A_0 C_0}\right)\\
&= 1 - \frac{2(A_0 + B_0 + C_0)}{2A_0 + B_0 \pm \sqrt{B_0^2 - 4A_0 C_0}},
\end{aligned}\tag{16}
$$

where the subscripts o, e refer to the ordinary and the extraordinary wave, respectively, and the zeroth order coefficients A_0, B_0, C_0 read immediately from Equations (7)

$$
\begin{aligned}
A_0 &= \varepsilon_{11}^{(0)}\sin^2\theta + \varepsilon_{33}^{(0)}\cos^2\theta,\\
-B_0 &= \varepsilon_{11}^{(0)}\varepsilon_{33}^{(0)}(1 + \cos^2\theta) + (\varepsilon_{11}^{(0)2} + \varepsilon_{12}^{(0)2})\sin^2\theta,\\
C_0 &= (\varepsilon_{11}^{(0)2} + \varepsilon_{12}^{(0)2})\varepsilon_{33}^{(0)}.
\end{aligned}\tag{17}
$$

Then, Equation (16) may be written more explicitly as

$$
n_{o,e}^{(0)2} = \frac{\substack{(\varepsilon_{11}^{(0)2} + \varepsilon_{12}^{(0)2})\sin^2\theta + \varepsilon_{11}^{(0)}\varepsilon_{33}^{(0)}(1 + \cos^2\theta)\\[4pt] \pm\sqrt{(\varepsilon_{11}^{(0)2} + \varepsilon_{12}^{(0)2} - \varepsilon_{11}^{(0)}\varepsilon_{33}^{(0)})^2\cdot\sin^4\theta - 4\varepsilon_{12}^{(0)2}\varepsilon_{33}^{(0)2}\cos^2\theta}}{2(\varepsilon_{11}^{(0)}\sin^2\theta + \varepsilon_{33}^{(0)}\cos^2\theta)}.\tag{18}
$$

This may include the ion dynamics via the cold dielectric coefficients (11).

The first-order thermal corrections to Equation (16) or (18) lead to

$$
\begin{aligned}
n_{o,e}^2 &= n_{o,e}^{(0)2}\left(1 - \sum_{+,-}\epsilon_\pm\,\frac{A_{1\pm}n_{o,e}^{(0)4} + B_{1\pm}n_{o,e}^{(0)2} + C_{1\pm}}{2A_0 n_{o,e}^{(0)2} + B_0}\right)\\
&= n_{o,e}^{(0)2}\left(1 \mp \sum_{+,-}\epsilon_\pm\,\frac{A_{1\pm}n_{o,e}^{(0)4} + B_{1\pm}n_{o,e}^{(0)2} + C_{1\pm}}{\sqrt{B_0^2 - 4A_0 C_0}}\right),
\end{aligned}\tag{19}
$$

where again the summation is to be taken for ions $(+)$ and electrons $(-)$.

Let us find the solution of Equation (14) for $|A_0| \ll 1$. The refractive index of the ordinary wave is obtained from Equations (16) and (19) in the form

$$n_o^2 \simeq n_o^{(0)2} \left\{ 1 - \sum_{+,-} \frac{\epsilon_\pm}{B_0} (A_{1\pm} n_o^{(0)4} + B_{1\pm} n_o^{(0)2} + C_{1\pm}) \right\},$$

$$n_o^{(0)2} \simeq - \frac{C_0}{B_0}, \qquad (20)$$

while the refractive index of the extraordinary wave becomes so large that the latter tends to couple to the plasma wave, i.e. $|n_{e,p}^2| \gg 1$. Then the refractive indices of the extraordinary and plasma waves may be written as

$$n_{e,p}^2 \simeq \frac{A_0}{2(\epsilon_+ A_{1+} + \epsilon_- A_{1-})} \left\{ -1 \pm \sqrt{1 - 4(\epsilon_+ A_{1+} + \epsilon_- A_{1-}) \frac{B_0}{A_0^2}} \right\}. \quad (21)$$

Further, for $1 \gg |A_0| \gg |\epsilon_+ A_{1+} + \epsilon_- A_{1-}|$, $|B_0| \gtrless |A_0|$ Equation (21) reduces to

$$n_e^2 \simeq - B_0/A_0 \qquad (22)$$

and

$$n_p^2 \simeq - \frac{A_0}{\epsilon_+ A_{1+} + \epsilon_- A_{1-}}, \qquad (23)$$

while for $|A_0| \ll |\epsilon_+ A_{1+} + \epsilon_- A_{1-}| \ll 1 \lessgtr |B_0|$ Equation (21) becomes

$$n_{e,p}^2 \simeq \pm \sqrt{- \frac{B_0}{\epsilon_+ A_{1+} + \epsilon_- A_{1-}}}. \qquad (24)$$

The phase velocity of the plasma waves introduced in Equations (21), (23), and (24) should be still large compared to the electron thermal velocity, since these waves were obtained under the quasi-cold conditions (8), (9), i.e. $|\mu_\pm| \ll 1$, $|z_{p\pm}| \gg 1$ $(p = 0, \pm 1, \pm 2, \cdots)$, particularly $|z_{0+}| \gg |z_{0-}| \gg 1$. Therefore the formulas are exact for the fast electron plasma wave far from cyclotron resonances. For a slower electron plasma wave and an ion plasma wave or close enough to the cyclotron frequencies, Landau and/or cyclotron damping become appreciable, i.e. $|z_{0\pm}| \gtrsim 1$ and/or $|z_{p\pm}| \gtrsim 1$ $(p \neq 0)$. This is the case for Čerenkov and/or cyclotron resonances and will be discussed in the later sections.

Furthermore, we have ignored the magnetosonic regime in which a part of the quasi-cold condition breaks down, i.e. (i) $|z_{0-}| \ll 1$ but still $|z_{0+}| \gg 1$, $|z_{p\pm}| \gg 1$ $(p \neq 0)$ or (ii) $|z_{0-}| \ll 1$ and even $|z_{0+}| \ll 1$ but still $|z_{p\pm}| \gg 1$ $(p \neq 0)$. The case (i) is readily realized above a height of 20000 to 30000 km, i.e. in the region of transition from the magnetosphere to the interplanetary space; the Alfvén velocity is small compared to the electron thermal velocity in this region, i.e. $v_A \ll v_{T-}$ (Equation (5)). As the altitude is further increased, it is more likely that the case (i) tends to the case (ii). Both cases will also be discussed later in some detail.

4. Magnetoionic and Electron Plasma Waves

In Section 2, it was indicated that the inclusion of the thermal or temperature effects is required under certain circumstances, particularly for top-side soundings, and that the normal collisional effects can be neglected most of the time for the outer ionosphere. Specifically, the top-side ionograms, for instance obtained from the swept radio-frequency sounder Alouette, demonstrate the appearance of the plasma waves in addition to the normal electromagnetic waves, ordinary and extraordinary. This indicates that the classical Appleton–Hartree equation should be generalized so as to include the thermal effects on plasma waves.

The same argument also applies to magnetoionic and plasma waves in the VLF range, for a precise analysis of whistler data and for investigating excitation mechanisms of non-ducted whistlers and VLF radiations which also have been observed by satellite-borne experiments. For this purpose an attempt is made in this section to generalize the classical Appleton–Hartree equation so as to meet these requirements over the HF and VLF ranges.

For the HF and VLF ranges where $\omega \gtrsim \omega_{c-}$ and $\omega_{c+} \ll \omega_h \lesssim \omega < \omega_{c-}$; $\omega_h =$ lower hybrid frequency $\simeq \sqrt{\omega_{c+} \omega_{c-}}$, ion contribution can be mostly neglected in all expressions in Section 3. Then the cold refractive indices (16) read

$$n_{o,e}^{(0)2} = 1 - \frac{2\varepsilon_{33}^{(0)}(1 - \varepsilon_{11}^{(0)})(1 - \varepsilon_{33}^{(0)})}{2\varepsilon_{33}^{(0)}(1 - \varepsilon_{11}^{(0)}) + (\varepsilon_{11}^{(0)} - \varepsilon_{33}^{(0)})\sin^2\theta \\ \pm \sqrt{(\varepsilon_{11}^{(0)} - \varepsilon_{33}^{(0)})^2 \sin^4\theta - 4\varepsilon_{33}^{(0)2}(1 - \varepsilon_{11}^{(0)})\cdot(\varepsilon_{11}^{(0)} - \varepsilon_{33}^{(0)})\cos^2\theta}}$$

$$(25)$$

where, referring to Equations (11),

$$\varepsilon_{11}^{(0)} = \varepsilon_{22}^{(0)} = 1 - \frac{\omega_{p-}^2}{\omega^2 - \omega_{c-}^2},$$

$$\varepsilon_{33}^{(0)} = 1 - \omega_{p-}^2/\omega^2,$$

$$\varepsilon_{12}^{(0)} = -\varepsilon_{21}^{(0)} = \frac{i\omega_{p-}^2 \omega_{c-}/\omega}{\omega^2 - \omega_{c-}^2} \tag{26}$$

and the relation

$$\varepsilon_{12}^{(0)2} = (1 - \varepsilon_{11}^{(0)})(\varepsilon_{11}^{(0)} - \varepsilon_{33}^{(0)}) \tag{27}$$

was employed. Equation (25) is an alternative form of the classical Appleton–Hartree equation and for the thermal corrections to this, only electrons contribute in Equation (19). We consider the two simplifications based on the relative magnitude of the two terms within the radical in Equation (25).

4.1. QUASI-PARALLEL PROPAGATION

One approximation is made by setting

$$|(\varepsilon_{11}^{(0)} - \varepsilon_{33}^{(0)})\sin^4\theta| \ll 4|\varepsilon_{33}^{(0)2}(1 - \varepsilon_{11}^{(0)})\cos^2\theta| \tag{28}$$

and yields

$$n_{o,e}^{(0)2} \simeq 1 - \frac{1 - \varepsilon_{33}^{(0)}}{1 \pm \sqrt{\dfrac{\varepsilon_{33}^{(0)} - \varepsilon_{11}^{(0)}}{1 - \varepsilon_{11}^{(0)}}} \cos\theta} = 1 - \frac{\omega_{p-}^2}{\omega(\omega \pm \omega_{c-}\cos\theta)}. \tag{29}$$

Then Equation (19) with the thermal corrections reduces to

$$n_{o,e}^2 = n_{o,e}^{(0)2} \left\{ 1 \mp \epsilon_- \frac{A_{1-}n_{o,e}^{(0)4} + B_{1-}n_{o,e}^{(0)2} + C_{1-}}{2\varepsilon_{33}^{(0)}\sqrt{(1 - \varepsilon_{11}^{(0)})(\varepsilon_{33}^{(0)} - \varepsilon_{11}^{(0)})}\cos\theta} \right\}, \tag{30}$$

where, referring to Equations (13) and (15),

$$\begin{aligned}
A_{1-} &\simeq \delta_{33-}\cos^2\theta, \\
-B_{1-} &\simeq (\varepsilon_{11}^{(0)}\delta_{33-} + \varepsilon_{33}^{(0)}\delta_{11-})(1 + \cos^2\theta), \\
C_{1-} &\simeq 2\varepsilon_{33}^{(0)}(\varepsilon_{11}^{(0)}\delta_{11-} + \varepsilon_{12}^{(0)}\delta_{12-}),
\end{aligned} \tag{31}$$

and

$$\delta_{11-} \simeq \delta_{22-} \simeq -\frac{\omega_{p-}^2\omega^2(\omega^2 + 3\omega_{c-}^2)}{(\omega^2 - \omega_{c-}^2)^3}\cos^2\theta,$$

$$\delta_{33-} \simeq -\frac{3\omega_{p-}^2}{\omega^2}\cos^2\theta,$$

$$\delta_{12-} = -\delta_{21-} \simeq \frac{i\omega_{p-}^2\omega_{c-}\omega(3\omega^2 + \omega_{c-}^2)}{(\omega^2 - \omega_{c-}^2)^3}\cos^2\theta. \tag{32}$$

With reference to 'whistlers', i.e. in the frequency range $\omega_h < \omega \lesssim \omega_{c-}$, only the extraordinary wave propagates, since $\omega_{p-}^2 \gg \omega_{c-}^2$ and the relation $|n_{o,e}^{(0)2}| \gg 1$ holds. Then, setting $\theta \simeq 0$ in Equations (29) to (32), we obtain the refractive index for the whistler mode with the thermal correction

$$n_e = \frac{\omega_{p-}}{\sqrt{\omega(\omega_{c-} - \omega)}}\left\{ 1 + \frac{\epsilon_-}{2}\cdot\frac{\omega_{p-}^2\omega}{(\omega_{c-} - \omega)^3} \right\}. \tag{33}$$

4.2. QUASI-PERPENDICULAR PROPAGATION

The other approximation for Equation (25) is made by setting

$$|(\varepsilon_{11}^{(0)} - \varepsilon_{33}^{(0)})\sin^4\theta| \gg 4|\varepsilon_{33}^{(0)2}(1 - \varepsilon_{11}^{(0)})\cos^2\theta| \tag{34}$$

and yields

$$n_o^{(0)2} \simeq \frac{\varepsilon_{33}^{(0)}}{1 - (1 - \varepsilon_{33}^{(0)})\cos^2\theta} = \frac{\omega^2 - \omega_{p-}^2}{\omega^2 - \omega_{p-}^2\cos^2\theta},$$

$$\begin{aligned}
n_e^{(0)2} &\simeq \frac{\varepsilon_{33}^{(0)2}(1 - \varepsilon_{11}^{(0)}) + (\varepsilon_{11}^{(0)} - \varepsilon_{33}^{(0)})\sin^2\theta}{\varepsilon_{33}^{(0)}(1 - \varepsilon_{11}^{(0)}) + (\varepsilon_{11}^{(0)} - \varepsilon_{33}^{(0)})\sin^2\theta} \\
&= \frac{(\omega^2 - \omega_{p-}^2)^2 - \omega_{c-}^2\omega^2\sin^2\theta}{\omega^2(\omega^2 - \omega_{p-}^2 - \omega_{c-}^2\sin^2\theta)}.
\end{aligned} \tag{35}$$

Then Equation (19) with the thermal corrections leads to

$$n^2_{o,e} = n^{(0)2}_{o,e} \left\{ 1 \mp \epsilon_- \frac{A_{1-}n^{(0)4}_{o,e} + B_{1-}n^{(0)2}_{o,e} + C_{1-}}{(\varepsilon^{(0)}_{33} - \varepsilon^{(0)}_{11})\sin^2\theta} \right\}, \tag{36}$$

where, referring to Equations (13) and (15),

$$A_{1-} \simeq \delta_{11-}\sin^2\theta,$$
$$-B_{1-} \simeq \varepsilon^{(0)}_{11}\delta_{33-} + \varepsilon^{(0)}_{33}\delta_{11-} + \{\varepsilon^{(0)}_{11}(\delta_{11-} + \delta_{22-}) + 2\varepsilon^{(0)}_{12}\delta_{12-}\}\sin^2\theta,$$
$$C_{1-} \simeq \varepsilon^{(0)}_{33}\{\varepsilon^{(0)}_{11}(\delta_{11-} + \delta_{22-}) + 2\varepsilon^{(0)}_{12}\delta_{12-}\}, \tag{37}$$

and

$$\delta_{11-} \simeq -\frac{3\omega^2_{p-}\omega^2}{(\omega^2 - \omega^2_{c-})(\omega^2 - 4\omega^2_{c-})}\sin^2\theta,$$

$$\delta_{22-} \simeq -\frac{\omega^2_{p-}(\omega^2 + 8\omega^2_{c-})}{(\omega^2 - \omega^2_{c-})(\omega^2 - 4\omega^2_{c-})}\sin^2\theta,$$

$$\delta_{12-} \simeq -\delta_{21-} \simeq \frac{i6\omega^2_{p-}\omega_{c-}\omega}{(\omega^2 - \omega^2_{c-})(\omega^2 - 4\omega^2_{c-})}\sin^2\theta,$$

$$\delta_{33-} \simeq -\frac{\omega^2_{p-}}{\omega^2 - \omega^2_{c-}}\sin^2\theta. \tag{38}$$

4.3. ELECTRON CYCLOTRON DAMPING: WEAK DAMPING

So far we have ignored all Landau and cyclotron damping under the quasi-cold conditions. When the wave frequency approaches the cyclotron frequencies, however, cyclotron damping becomes appreciable even under the quasi-cold conditions, since the magnitude of a cyclotron resonance parameter $|z_{p\pm}|$ ($p \neq 0$) approaches unity from above. We now consider electron cyclotron damping in the range of frequencies not so close to the electron cyclotron frequency and retain the largest cyclotron damping term in the complex error functions of general formulas. Then the cyclotron damping corrections to the dielectric coefficients (10) yield

$$\tilde{\varepsilon} = \tilde{\varepsilon}^{(0)} + n^2\epsilon_-\tilde{\delta}_- + i\frac{\tilde{\gamma}_-}{n\sqrt{\epsilon_-}}, \tag{39}$$

where

$$\tilde{\gamma}_- = \delta_{-1} \begin{Vmatrix} 1 & -i & 0 \\ i & 1 & 0 \\ 0 & 0 & 0 \end{Vmatrix} + \kappa_{-1} \begin{Vmatrix} 0 & 0 & 1 \\ 0 & 0 & i \\ 1 & -i & 0 \end{Vmatrix},$$

$$\delta_{-1} = \frac{1}{2}\sqrt{\frac{\pi}{2}} \cdot \frac{\omega^2_{p-}}{\omega^2\cos\theta} e^{-z^2_{1-}}, \quad \kappa_{-1} = -\left(1 - \frac{\omega}{\omega_{c-}}\right)\delta_{-1}\tan\theta, \tag{40}$$

and the real part in Equation (39) is defined in Equations (10) to (13). Then the dispersion relation (6) takes on the form, making the damping corrections to Equation (14),

$$\epsilon_-A_{1-}n^7 + (A_0 + \epsilon_-B_{1-})n^5 + (B_0 + \epsilon_-C_{1-})n^3 + C_0n$$
$$+ i\epsilon^{-1/2}\delta_{-1}(A_{-1-}n^4 + B_{-1-}n^2 + C_{-1-}) = 0, \tag{41}$$

where

$$A_{-1-} = -\left(1 - \frac{2\omega}{\omega_{c-}}\right)\sin^2\theta,$$

$$-B_{-1-} = \varepsilon_{33}^{(0)}(1 + \cos^2\theta) + \frac{2\omega}{\omega_{c-}}(\varepsilon_{11}^{(0)} - i\varepsilon_{12}^{(0)})\sin^2\theta,$$

$$C_{-1-} = 2\varepsilon_{33}^{(0)}(\varepsilon_{11}^{(0)} - i\varepsilon_{12}^{(0)}). \tag{42}$$

Since the neglect of ion contribution in Equation (19) forms the solutions of Equation (41) for $\delta_{-1}=0$, the damping correction yields

$$n_{o,e}^2 \simeq n_{o,e}^{(0)2}\left(1 \mp \varepsilon_- \frac{A_{1-}n_{o,e}^{(0)4} + B_{1-}n_{o,e}^{(0)2} + C_{1-}}{\sqrt{B_0^2 - 4A_0C_0}}\right.$$

$$\left.\mp i\frac{\varepsilon_-^{-1/2}\delta_{-1}}{n_{o,e}^{(0)3}} \cdot \frac{A_{-1-}n_{o,e}^{(0)4} + B_{-1-}n_{o,e}^{(0)2} + C_{-1-}}{\sqrt{B_0^2 - 4A_0C_0}}\right), \tag{43}$$

where the imaginary part represents spatial cyclotron damping and becomes large when the wave frequency approaches the electron cyclotron frequency.

With particular reference to 'whistlers', we again set $\omega_h < \omega \lesssim \omega_{c-}$, $\theta \simeq 0$ in Equations (39) to (43) and obtain the cyclotron damping correction to Equation (33)

$$n_e \simeq \frac{\omega_{p-}}{\sqrt{\omega(\omega_{c-} - \omega)}}\left(1 + \frac{\varepsilon_-}{2} \cdot \frac{\omega_{p-}^2 - \omega}{(\omega_{c-} - \omega)^3}\right.$$

$$\left.+ \frac{i}{2}\sqrt{\frac{\pi}{2\varepsilon_-}} \frac{\sqrt{(\omega_{c-} - \omega)^3}}{\omega_{p-}\sqrt{\omega}} e^{-z_{1-}^2}\right). \tag{44}$$

4.4. ELECTRON CYCLOTRON DAMPING: ELECTRON CYCLOTRON RESONANCE

When the wave frequency becomes close enough to the electron cyclotron frequency, one of the quasi-cold conditions no longer holds, i.e. $|z_{-1-}| \lesssim 1$. Then the treatment in the preceding subsection not only fails but also expressions for the dielectric coefficients are radically altered. To obtain new formulas we must employ the convergent series expansion instead of the asymptotic expansion for the complex error functions concerned. We thus obtain the dielectric coefficients for the case of the electron cyclotron resonance ($|z_{-1-}| \ll 1$)

$$\tilde{\varepsilon} = \tilde{\varepsilon}^{(0)} + i\frac{\tilde{\gamma}_-}{n\sqrt{\varepsilon_-}}, \tag{45}$$

where

$$\tilde{\varepsilon}_{11}^{(0)} \simeq \tilde{\varepsilon}_{22}^{(0)} \simeq 1 - \frac{\omega_{p-}^2}{2\omega(\omega + \omega_{c-})}, \quad \tilde{\varepsilon}_{33}^{(0)} \simeq \varepsilon_{33}^{(0)} = 1 - \frac{\omega_{p-}^2}{\omega^2},$$

$$\tilde{\varepsilon}_{12}^{(0)} \simeq -\tilde{\varepsilon}_{21}^{(0)} \simeq -\frac{i\omega_{p-}^2}{2\omega(\omega + \omega_{c-})}, \quad \tilde{\varepsilon}_{13}^{(0)} = \tilde{\varepsilon}_{31}^{(0)} \simeq \frac{\omega_{p-}^2 \tan\theta}{2\omega_{c-}\omega},$$

$$\tilde{\varepsilon}_{23}^{(0)} = -\tilde{\varepsilon}_{32}^{(0)} \simeq \frac{i\omega_{p-}^2 \tan\theta}{2\omega_{c-}\omega}, \tag{46}$$

$$\tilde{\gamma} = \delta_{-1} \begin{Vmatrix} 1 & -i & 0 \\ i & 1 & 0 \\ 0 & 0 & 0 \end{Vmatrix}, \qquad \delta_{-1} = \frac{1}{2} \sqrt{\frac{\pi}{2}} \cdot \frac{\omega_{p-}^2}{\omega^2 \cos\theta} e^{-z^2_{-1-}}. \qquad (47)$$

Then the dispersion relation (6) takes on the form

$$A_0 n^5 + B_0 n^3 + C_0 n + i\epsilon_-^{-1/2}\delta_{-1}(A_{-1-}n^4 + B_{-1-}n^2 + C_{-1-}) = 0, \quad (48)$$

where

$$\begin{aligned} A_{-1-} &= \sin^2\theta, \\ -B_{-1-} &= \varepsilon_{33}^{(0)}(1 + \cos^2\theta) + 2(\hat{\varepsilon}_{11}^{(0)} - i\hat{\varepsilon}_{12}^{(0)})\sin^2\theta, \\ C_{-1-} &= 2\varepsilon_{33}^{(0)}(\hat{\varepsilon}_{11}^{(0)} - i\hat{\varepsilon}_{12}^{(0)}), \end{aligned} \qquad (49)$$

and A_0, B_0, C_0 are defined by setting $\hat{\varepsilon}_{ik}^{(0)}$ instead of ε_{ik} in Equations (7). It should be noted, therefore, that the dielectric coefficients for the case of the electron cyclotron resonance must be taken from Equations (46), which are quite different from those for a quasi-cold plasma.

Retaining only terms proportional to $\varepsilon_-^{-1/2}$, the solution of Equation (48) may be written as

$$n_{o,e}^{(0)2} = 1 - \frac{2(A_{-1-} + B_{-1-} + C_{-1-})}{2A_{-1-} + B_{-1-} \pm \sqrt{B_{-1-}^2 - 4A_{-1-}C_{-1-}}}. \qquad (50)$$

In the next approximation we find

$$n_{o,e}^2 = n_{o,e}^{(0)2}\left(1 \pm i\varepsilon_-^{1/2} \frac{A_0 n_{o,e}^{(0)4} + B_0 n_{o,e}^{(0)2} + C_0}{\delta_{-1} n_{o,e}^{(0)}\sqrt{B_{-1-}^2 - 4A_{-1-}C_{-1-}}}\right)$$
$$\text{for} \quad \epsilon_-^{-1/2}|\delta_{-1}|\sin^2\theta \gg 1. \quad (51)$$

Thus the electromagnetic waves are damped near the electron cyclotron frequency and the magnitude of the damping coefficient is of order $\epsilon_-^{1/2}$.

4.5. ELECTRON LANDAU DAMPING

In Section 3, a fast electron plasma and a slow electromagnetic wave were introduced under the quasi-cold conditions neglecting Landau damping ($|z_{0-}| \gg 1$). When the plasma wave is slowed and approaches the electron thermal velocity, its interaction with the thermal electrons increases and gives rise to Landau damping ($|z_{0-}| \gtrsim 1$), i.e. the condition of Čerenkov resonance is almost attained. This, however, does not change the treatment of the plasma waves in Section 3 so much but only makes a damping correction. Basically, expressions for the dielectric coefficients (10), (11), (13) are not altered except that a correction must be made for the dielectric coefficient ε_{33} in the form

$$\varepsilon_{33} = \varepsilon_{33}^{(0)} + n^2(\epsilon_+\delta_{33+} + \epsilon_-\delta_{33-}) + i\sqrt{\frac{\pi}{2}} \cdot \frac{\omega_{p-}^2 e^{-z^2_{0-}}}{\omega^2 n^3 \epsilon_-^{3/2}\cos^3\theta}. \qquad (52)$$

The first two terms are the quasi-cold terms defined in Equations (10) to (13), and the last imaginary term represents a new inclusion of electron Landau damping which has a maximum at $z_0 = \sqrt{1.5} = 1.225$.

Then the addition of electron Landau damping into Equation (14) yields

$$(\epsilon_+ A_{1+} + \epsilon_- A_{1-})n^5 + (A_0 + \epsilon_+ B_{1+} + \epsilon_- B_{1-})n^3$$
$$+ (B_0 + \epsilon_+ C_{1+} + \epsilon_- C_{1-})n + i\epsilon_-^{-3/2}\delta_{0-} = 0, \quad (53)$$

where

$$\delta_{0-} = \sqrt{\frac{\pi}{2}} \frac{\omega_{p-}^2}{\omega^2} \cdot \frac{e^{-z^2 0-}}{\cos\theta}. \quad (54)$$

Since one of the solutions of Equation (53) for $\delta_{0-} = 0$ forms the refractive indices of the coupled extraordinary and electron plasma waves for zero Landau damping, the damping correction for Equation (21) leads to

$$n_{e,p}^2 = n_{e,p}^{(0)2}\left\{1 - i\frac{\epsilon_-^{-3/2}\delta_{0-}}{n_{e,p}^{(0)3}(2n_{e,p}^{(0)2}\sum_{+,-}\epsilon_\pm A_{1\pm} + A_0)}\right\}$$

$$= n_{e,p}^{(0)2}\left\{1 + i\frac{\epsilon_-^{-3/2}\delta_{0-}}{n_{e,p}^{(0)}(A_0 n_{e,p}^{(0)2} + 2B_0)}\right\}, \quad (55)$$

where $n_{e,p}^{(0)}$ indicates the refractive indices of the extraordinary and plasma waves for zero Landau damping and is defined by Equation (21).

For $|A_0| \gg |\epsilon_+ A_{1+} + \epsilon_- A_{1-}|$, $|B_0| \gtrless |A_0|$ in which the plasma wave becomes slower but the extraordinary wave faster, the damping correction for the plasma waves yields, referring to Equations (23) and (55),

$$n_p^2 = -\frac{A_0}{\epsilon_+ A_{1+} + \epsilon_- A_{1-}}\left\{1 + i\frac{\delta_{0-}}{A_0}\left(-\frac{A_{1-} + \xi\eta A_{1+}}{A_0}\right)^{3/2}\right\}, \quad (56)$$

where $\xi = T_+/T_-$, $\eta = m_-/m_+$, while for $|A_0| \ll |\epsilon_+ A_{1+} + \epsilon_- A_{1-}| \gtrless |B_0|$ which gives plasma resonances such as the hybrid resonances or Z-infinity, we obtain the damping correction for Equation (24) from Equation (55) in the form

$$n_{ep}^2 = \sqrt{-\frac{B_0}{\epsilon_+ A_{1+} + \epsilon_- A_{1-}}}\left\{1 + i\frac{\epsilon_-^{-5/4}\delta_{0-}}{2B_0}\left(-\frac{A_{1-} + \xi\eta A_{1+}}{B_0}\right)^{1/4}\right\}. \quad (57)$$

The subscript ep indicates that the plasma wave is coupled with the extraordinary wave.

4.6. COUPLED ELECTRON CYCLOTRON AND WEAK ČERENKOV RESONANCES

For the case when both electron cyclotron and Čerenkov resonances almost overlap, i.e. when the longitudinal electron Landau damping predominates ($|z_0| \gtrsim 1$) in the immediate vicinity of the electron cyclotron frequency ($|z_{-1-}| \ll 1$), Equation (6) takes on the form, referring to Equations (48) and (53) and taking into account the largest terms of both cyclotron and Landau damping,

$$\epsilon_- A_{1-}n^7 + A_0 n^5 + B_0 n^3 + C_0$$
$$+ i\{\epsilon_-^{-1/2}\delta_{-1}(A_{-1-}n^4 + B_{-1-}n^2 + C_{-1-}) + \epsilon_-^{-3/2}\delta_{0-}n^2\} = 0, \quad (58)$$

where the terms with δ_{-1} and δ_{0-} refer to cyclotron and Landau damping, respectively. When cyclotron damping is strong enough compared to electron Landau damping, the Landau damping correction to Equation (51) yields

$$
n_{o,e}^2 = n_{o,e}^{(0)2}\left(1 \pm i\epsilon_-^{1/2}\,\frac{A_0 n_{o,e}^{(0)4} + B_0 n_{o,e}^{(0)2} + C_0}{\delta_{-1} n_{o,e}^{(0)}\sqrt{B_{-1-}^2 - 4A_{-1-}C_{-1-}}}\right.
$$
$$
\left.\pm\,\frac{i_-\epsilon_-^{3/2}A_{1-}n^{(0)5}}{\cdot\delta_{-1}\sqrt{B_{-1-}^2 - 4A_{-1-}C_{-1-}}\pm\epsilon_-^{-1}\delta_{0-}}\right) \quad \text{for}\quad \epsilon_-^{-1/2}|\delta_{-1}|\sin^2\theta \gg 1.
$$
(59)

The first two terms form the solution for cyclotron resonances without Landau damping which was obtained in Equation (51), and the last term represents the effect of electron Landau damping.

4.7. ELECTRON LANDAU AND WEAK CYCLOTRON DAMPING

For the case when electron Landau damping predominates ($|z_{0-}| \gtrsim 1$) near but not too near the electron cyclotron frequency ($|z_{-1-}| > 1$), the electron plasma and slow electromagnetic waves are affected by weak cyclotron damping. Then Equation (6) takes on the form, referring to Equations (41) and (53) and taking into account the largest terms of electron Landau and cyclotron damping,

$$
\epsilon_- A_{1-} n^5 + (A_0 + \epsilon_- B_{1-})n^3 + (B_0 + \epsilon_- C_{1-})n
$$
$$
+ i\epsilon_-^{-3/2}\{\delta_{0-} + \epsilon_-\delta_{-1}(A_{-1-}n^2 + B_{-1-})\} = 0, \quad (60)
$$

where the terms with δ_{0-} and δ_{-1} refer to Landau and cyclotron damping, respectively, and ion contributions were all neglected. Since the solution of Equation (60) for $\delta_{-1}=0$ is given by Equation (55), the cyclotron damping correction for Equation (55) yields

$$
n_{e,p}^2 = n_{e,p}^{(0)2}\left\{1 - i\epsilon_-^{-3/2}\frac{\delta_{0-} + \epsilon_-\delta_{-1}(A_{-1-}n_{e,p}^{(0)2} + B_{-1-})}{n_{e,p}^{(0)3}(2\epsilon_- A_{1-}n_{e,p}^{(0)2} + A_0)}\right\}
$$
$$
= n_{e,p}^{(0)2}\left\{1 + i\epsilon_-^{-3/2}\frac{\delta_{0-} + \epsilon_-\delta_{-1}(A_{-1-}n_{e,p}^{(0)2} + B_{-1-})}{n_{e,p}^{(0)}(A_0 n_{e,p}^{(0)2} + 2B_0)}\right\}.
$$
(61)

The imaginary part expresses the effects of both electron Landau and cyclotron damping.

For $|A_0| \gg |\epsilon_- A_{1-}|$, $|B_0| \gtrless |A_0|$, we obtain the cyclotron damping correction for Equation (56) in the form

$$
n_p^2 = -\frac{A_0}{\epsilon_- A_{1-}}\left\{1 + i\frac{\delta_{0-}}{A_0}\left(1 - \frac{\delta_{-1}}{\delta_{0-}}\cdot\frac{A_0 A_{-1-}}{A_{1-}}\right)\left(-\frac{A_{1-}}{A_0}\right)^{3/2}\right\}, \quad (62)
$$

while for $|A_0| \ll |\epsilon_- A_{1-}| \ll 1 \lesssim |B_0|$, the cyclotron damping correction for Equation (57) takes on the form

$$
n_{ep}^2 = \sqrt{-\frac{B_0}{\epsilon_- A_{1-}}}\left\{1 + i\frac{\epsilon_-^{-5/4}\delta_{0-}}{2B_0}\left(1 - \frac{\delta_{-1}}{\delta_{0-}}\cdot\frac{A_0 A_{-1-}}{A_{1-}}\right)\left(-\frac{A_{1-}}{B_0}\right)^{1/4}\right\}.
$$
(63)

4.8. Second harmonic electron cyclotron resonance

When the wave frequency falls close enough to the second harmonic of the electron cyclotron frequency, one of the quasi-cold conditions no longer holds, i.e. $|z_{-2-}| \ll 1$. Then, expressions for the dielectric coefficients (10), (11), (13) are altered and the same procedure as was carried out for the vicinity of the cyclotron frequency yields

$$\tilde{\varepsilon} = \tilde{\varepsilon}^{(0)} + in\sqrt{\varepsilon_-}\,\tilde{\gamma}, \tag{64}$$

where

$$\tilde{\gamma} = \delta_{-2} \begin{Vmatrix} 1 & -i & 0 \\ i & 1 & 0 \\ 0 & 0 & 0 \end{Vmatrix},$$

$$\delta_{-2} = \frac{1}{2}\sqrt{\frac{\pi}{2}} \cdot \frac{\omega_{p-}^2}{\omega_{c-}^2} \cdot \frac{\sin^2\theta}{\cos\theta}\, e^{-z^2_{-2-}}, \tag{65}$$

and $\tilde{\varepsilon}^{(0)}$ represents the cold dielectric tensor defined in Equation (26). The dispersion Equation (6) then takes on the form

$$A_0 n^4 + B_0 n^2 + C_0 + i\varepsilon_-^{1/2}\delta_{-2}n\,(A_{-2-}n^4 + B_{-2-}n^2 + C_{-2-}) = 0, \tag{66}$$

where

$$A_{-2-} = \sin^2\theta,$$
$$-B_{-2-} = \varepsilon_{33}^{(0)}(1 + \cos^2\theta) + 2(\varepsilon_{11}^{(0)} - i\varepsilon_{12}^{(0)})\sin^2\theta,$$
$$C_{-2-} = 2\varepsilon_{33}^{(0)}(\varepsilon_{11}^{(0)} - i\varepsilon_{12}^{(0)}), \tag{67}$$

and the coefficients A_0, B_0, C_0 are defined in Equations (17). Then, the solution of Equation (66) may be written as

$$n_{o,e}^2 = n_{o,e}^{(0)2}\left\{1 \mp i\varepsilon^{1/2}\frac{\delta_{-2}(A_{-2-}n_{o,e}^{(0)4} + B_{-2-}n_{o,e}^{(0)2} + C_{-2-})}{n_{o,e}^{(0)}\sqrt{B_0^2 - 4A_0 C_0}}\right\}, \tag{68}$$

where $n_{o,e}^{(0)2}$ is defined in Equation (16). Thus the electromagnetic waves are damped at the second harmonic of the electron cyclotron frequency. The magnitude of damping is of order $\varepsilon^{1/2}$ and is equal to the order of cyclotron damping at the electron cyclotron frequency for oblique propagation.

4.9. Parallel propagation

For the case of exactly parallel propagation ($\theta = 0$), the rigorous treatment of a hot plasma is possible and the dispersion equation may simply be written as

$$n^2 = \tfrac{1}{2}(\varepsilon_{11} + \varepsilon_{22}) \mp i\varepsilon_{12}, \quad \varepsilon_{33} = 0, \tag{69}$$

which refer to the refractive indices of the ordinary, extraordinary, and plasma waves, respectively.

When electron cyclotron damping is taken into account, the refractive indices of the electromagnetic waves in Equation (69) may be written, referring to Equations (39) and (45) to (47), in the cubic form

$$n^3 - n_{o,e}^{(0)2}n - i\frac{2\delta_{-1}}{\sqrt{\epsilon_-}} = 0, \qquad n_{o,e}^{(0)2} = \varepsilon_{11}^{(0)} \mp i\varepsilon_{12}^{(0)}, \tag{70}$$

where δ_{-1} is defined in Equation (41); $\varepsilon_{11}^{(0)}$, $\varepsilon_{12}^{(0)}$ are defined in Equations (26) for frequencies not too close to the electron cyclotron frequency ($|z_{-1-}| \gg 1$), and in Equation (46) for its immediate vicinity ($|z_{-1-}| \lesssim 1$). When the wave frequency approaches the electron cyclotron frequency, the cyclotron damping term in Equation (70) gradually increases and becomes very large close enough to this frequency. Then the refractive indices of the ordinary and extraordinary waves deviate from their cold quantities $n_{o,e}^{(0)}$, causing a large discrepancy in the extraordinary wave at the cyclotron resonance.

The solutions of Equation (70) may be written, retaining only the largest terms in ϵ_-, as

$$\left.\begin{array}{c} n_o \\ n_e \\ n_b \end{array}\right\} = 2^{1/3}\epsilon_-^{-1/6}\delta_{-1}^{1/3} \left\{\begin{array}{c} e^{i3\pi/2} \\ e^{i\pi/6} \\ e^{i5\pi/6}. \end{array}\right. \tag{71}$$

While the ordinary wave does not propagate, the extraordinary wave is strongly damped near the electron cyclotron frequency. The magnitude of damping ($\mathrm{Im}\,n$) is of order $\epsilon_-^{-1/6}$ and is equal to the order of the real part of the refractive index ($\mathrm{Re}\,n$), provided that $|z_{-1-}| \ll 1$.

4.10. PERPENDICULAR PROPAGATION

For the case of exactly perpendicular propagation ($\theta = \pi/2$), there is no Landau nor cyclotron damping and the dispersion Equation (6) is simply written, referring to Equations (10) to (13), as

$$n^2 = \varepsilon_{33}, \qquad \varepsilon_{11}n^2 = \varepsilon_{11}\varepsilon_{22} + \varepsilon_{12}^2, \tag{72}$$

where the first and second relations define the refractive indices of the ordinary and extraordinary waves coupled with the plasma waves, respectively.

Referring to Equations (10) to (13) the first of Equations (72) yields, to the first order temperature corrections,

$$n_o^2 = \frac{n_o^{(0)2}}{1 - \epsilon_-\delta_{33-}} = \frac{n_o^{(0)2}(\omega^2 - \omega_{c-}^2)}{\omega^2 - \omega_{c-}^2 + \epsilon_-\omega_{p-}^2}. \tag{73}$$

This indicates that the ordinary wave has an electron cyclotron resonance close to but lower than the electron cyclotron frequency.

Referring again to Equations (10) to (13), the second of Equations (72) may be written, to order ϵ_-n^2, as

$$\epsilon_-\delta_{11-}n_e^4 + \{\varepsilon_{11}^{(0)} - \epsilon_-(\varepsilon_{11}'^{(0)}\delta_{22-}' + \varepsilon_{22}'^{(0)}\delta_{11-}')\}n_e^2 - \varepsilon_{11}'^{(0)}\varepsilon_{22}'^{(0)} = 0, \tag{74}$$

where

$$\left.\begin{matrix} \varepsilon_{11}^{\prime(0)} \\ \varepsilon_{22}^{\prime(0)} \end{matrix}\right\} = 1 - \frac{\omega_{p-}^2}{\omega(\omega \pm \omega_{c-})}, \qquad \left.\begin{matrix} \delta_{11-}^{\prime} \\ \delta_{22-}^{\prime} \end{matrix}\right\} = \tfrac{1}{2}(\delta_{11-} + \delta_{22-}) \mp i\delta_{12-}, \qquad (75)$$

and the prime refers to circularly polarized waves. Then, the zeroth order refractive index for the extraordinary wave may be written as

$$n_{e}^{(0)2} = \frac{\varepsilon_{11}^{\prime(0)}\varepsilon_{22}^{\prime(0)}}{\varepsilon_{11}^{(0)}} = \frac{(\omega^2 - \omega_{p-}^2)^2 - \omega_{c-}^2\omega^2}{\omega^2(\omega^2 - \omega_{p-}^2 - \omega_{c-}^2)} \qquad (76)$$

and the first order correction leads to

$$n_e^2 = n_e^{(0)2}\left\{1 + \frac{\epsilon_-}{\varepsilon_{11}^{(0)}}(\varepsilon_{11}^{\prime(0)}\delta_{22-}^{\prime} + \varepsilon_{22}^{\prime(0)}\delta_{11-}^{\prime}\right\}. \qquad (77)$$

The other solution of Equation (74) may be written, under the condition $|\varepsilon_{11}^{(0)}| \gg |\epsilon_-\delta_{11-}| \neq 0$, and referring to Equations (10) to (13), as

$$n_{ep}^2 = -\frac{\varepsilon_{11}^{(0)}}{\epsilon_-\delta_{11-}} = \frac{(\omega^2 - \omega_{p-}^2 - \omega_{c-}^2)(\omega^2 - 4\omega_{c-}^2)}{3\epsilon_-\omega_{p-}^2\omega^2}. \qquad (78)$$

The subscript ep indicates that this plasma wave is coupled with the extraordinary wave.

The upper and lower hybrid resonances ω_H, ω_h are found by setting $\varepsilon_{11}^{(0)}=0$, and we have the relationship between both resonance frequencies

$$\omega_H^2 + \omega_h^2 = \omega_{p+}^2 + \omega_{p-}^2 + \omega_{c+}^2 + \omega_{c-}^2,$$
$$\omega_H^2\omega_h^2 = \omega_{c+}\omega_{c-}(\omega_{p+}^2 + \omega_{p-}^2 + \omega_{c+}\omega_{c-}). \qquad (79)$$

In the immediate vicinity of these frequencies, we have, instead of Equations (77) and (78),

$$n_{ep}^2 = \sqrt{\frac{\varepsilon_{11}^{\prime(0)}\varepsilon_{22}^{\prime(0)}}{\epsilon_+\delta_{11+} + \epsilon_-\delta_{11-}}}. \qquad (80)$$

It should be noted that for the lower hybrid resonance the ion contribution must be taken into account in Equation (80).

5. Magnetohydrodynamic and Ion Plasma Waves

In the preceding section, we have treated the magnetoionic and electron plasma waves in the VLF range where the ion contribution has been largely ignored. Their lower limit has been taken tentatively to be the lower hybrid frequency. However, the ion plasma wave is possible in a non-isothermal plasma around the ion plasma frequency which becomes higher than the lower hybrid frequency beyond a height of about 10^4 km. The appearance of the lower hybrid resonance itself is already a manifestation of the ion effects, and well below this frequency the direct contribution of electrons rapidly diminishes except when electron Landau damping predominates, i.e. for the case of $|z_{0-}| \approx 1$.

As is well known, 'magnetohydrodynamics' is based on concepts of hydrodynamics familiar to fluid mechanicists, extended to compressible fluids in a magnetic field. However, the theory is still insufficient from the wave point of view in the sense that it restricts the problem to a range of frequencies much below the ion cyclotron frequency, i.e. $\omega \ll \omega_{c+}$. Recent observations of micropulsations and ELF radiations obtained from spectrograms with frequency-time display, particularly satellite observations of the so-called 'ion whistlers', indicate the necessity for complete inclusion of ion cyclotron waves in the magnetohydrodynamic waves. Furthermore, the Alfvén waves present a wide variety of modes or branches owing to velocities ranging from below the ion thermal velocity to far beyond the electron thermal velocity. This is because the Alfvén speed itself is highly dependent on altitude, and coupling of magnetosonic waves to ion plasma waves is particularly favored above a height of 20000 to 30000 km. This section therefore attempts a unified treatment of all kinds of waves in the ELF range.

For the ELF range where $\omega \ll \omega_{c+}$, $0 < \omega \lesssim \omega_{c+}$, or $\omega_{c+} < \omega \lesssim \omega_h \simeq \sqrt{(\omega_{c+} \omega_{c-})}$, expressions for the cold dielectric coefficients (11) may be simplified, under the quasi-cold conditions (8), (9), as

$$\varepsilon_{11}^{(0)} = \varepsilon_{22}^{(0)} = 1 - \frac{\omega_{p+}^2}{\omega^2 - \omega_{c+}^2}\left(1 - \frac{\omega^2}{\omega_{c+}\omega_{c-}}\right),$$

$$\varepsilon_{12}^{(0)} = -\varepsilon_{21}^{(0)} = -\frac{i\omega_{p+}^2\omega}{\omega_{c+}(\omega^2 - \omega_{c+}^2)},$$

$$\varepsilon_{33}^{(0)} = 1 - \frac{\omega_{p+}^2 + \omega_{p-}^2}{\omega^2} \simeq -\frac{\omega_{p-}^2}{\omega^2}. \tag{81}$$

For the thermal corrections in Equations (13), ion and electron contributions are comparable near the lower hybrid frequency $\omega_h \simeq \sqrt{(\omega_{c+} \omega_{c-})}$, although the ion contribution predominates near and below the ion cyclotron frequency.

Except for the immediate vicinity of the ion cyclotron frequency, i.e. under the condition

$$|1 - \omega_{c+}/\omega| \gg \eta = m_-/m_+, \tag{82}$$

the additional simplifying conditions follow as

$$|\varepsilon_{11}^{(0)}/\varepsilon_{33}^{(0)}| \ll 1, \quad |\varepsilon_{12}^{(0)}/\varepsilon_{33}^{(0)}| \ll 1, \quad |\varepsilon_{12}^{(0)2}/(\varepsilon_{11}^{(0)}\varepsilon_{33}^{(0)})| \ll 1, \tag{83}$$

and thus a simpler expression for Equations (17) is obtained, i.e.

$$A_0 = \varepsilon_{33}^{(0)} \cos^2\theta,$$

$$-B_0 = \varepsilon_{11}^{(0)}\varepsilon_{33}^{(0)}(1 + \cos^2\theta),$$

$$C_0 = \varepsilon_{33}^{(0)}(\varepsilon_{11}^{(0)2} + \varepsilon_{12}^{(0)2}). \tag{84}$$

Then the cold refractive indices, Equation (16) or (18), may be rewritten as

$$n_{o,e}^{(0)2} = \frac{1}{2\cos^2\theta}\{\varepsilon_{11}^{(0)}(1 + \cos^2\theta)$$

$$\pm \sqrt{\varepsilon_{11}^{(0)2}(1 + \cos^2\theta)^2 - 4(\varepsilon_{11}^{(0)2} + \varepsilon_{12}^{(0)2})\cos^2\theta}\}. \tag{85}$$

It should be noted that the zeroth order refractive indices are determined by the transverse dielectric coefficients $\varepsilon_{11}^{(0)}$, $\varepsilon_{12}^{(0)}$ only, since Equation (85) no longer contains the longitudinal dielectric constant $\varepsilon_{33}^{(0)}$.

With regard to the terminology for the refractive indices defined in Equation (85) for the ELF range, the ordinary wave with left-hand circular polarization is often referred to as the pure or the slow Alfvén wave, while the extraordinary wave with right-hand circular polarization is referred to as the modified or the fast Alfvén wave.

Except for the angles close to $\pi/2$, i.e. under the condition

$$\cos^2\theta \gg \eta, \tag{86}$$

the electron thermal corrections in Equations (13) may be simplified as

$$\delta_{11-} = \frac{3\omega_{p-}^2 - \omega^2}{\omega_{c-}^4}(\cos^2\theta - \tfrac{1}{4}\sin^2\theta),$$

$$\delta_{22-} = \frac{3\omega_{p-}^2 - \omega^2}{\omega_{c-}^4}\left(\cos^2\theta - \frac{2}{3}\cdot\frac{\omega_{c-}^2}{\omega^2}\sin^2\theta\right),$$

$$\delta_{33-} = -\frac{3\omega_{p-}^2}{\omega^2}\cos^2\theta,$$

$$\delta_{12-} = -\delta_{21-} = -i\frac{\omega_{p-}^2 - \omega}{\omega_{c-}^3}(\cos^2\theta - \tfrac{3}{2}\sin^2\theta),$$

$$\delta_{13-} = \delta_{31-} = -\frac{\omega_{p-}^2 - \omega^2}{\omega_{c-}^4}\sin 2\theta,$$

$$\delta_{23-} = -\delta_{32-} = i\frac{\omega_{p-}^2}{2\omega_{c-}\omega}\sin 2\theta, \tag{87}$$

with the relation

$$|\delta_{ik-}/\delta_{33-}| \ll 1 \quad (i = k \neq 3). \tag{88}$$

Then, the coefficients of the electron thermal corrections in Equation (15) reduce to

$$A_{1-} = \delta_{33-}\cos^2\theta,$$
$$-B_{1-} = \varepsilon_{11}^{(0)}\delta_{33-}(1+\cos^2\theta) + \varepsilon_{33}^{(0)}(\delta_{11-}+\delta_{22-}\cos^2\theta)$$
$$+ 2\varepsilon_{12}^{(0)}(\delta_{12-}\sin\theta - \delta_{23-}\cos\theta)\sin\theta,$$
$$C_{1-} = \varepsilon_{33}^{(0)}\{\varepsilon_{11}^{(0)}(\delta_{11-}+\delta_{22-}) + 2\varepsilon_{12}^{(0)}\delta_{12-}\}. \tag{89}$$

5.1. ION CYCLOTRON DAMPING: WEAK DAMPING

In the previous subsections, we treated the electron cyclotron damping near the electron cyclotron frequency. Similarly, when the wave frequency approaches the ion cyclotron frequency, ion cyclotron damping becomes appreciable even under the quasi-cold conditions, since the magnitude of the ion cyclotron resonance parameter $|z_{1+}|$ decreases to near unity. We now consider ion cyclotron damping in the range of frequencies not so close to the ion cyclotron frequency $(\omega \lesssim \omega_{c+})$ and retain the largest cyclotron damping terms in the complex error functions of general formulas.

Then the ion cyclotron damping correction to the dielectric coefficients (10) yields, under the condition (82),

$$\tilde{\varepsilon} = \tilde{\varepsilon}^{(0)} + n^2(\epsilon_+ \tilde{\delta}_+ + \epsilon_- \tilde{\delta}_-) + i\,\frac{\tilde{\gamma}_+}{n\sqrt{\epsilon_+}}, \tag{90}$$

where

$$\tilde{\gamma}_+ = \delta_1 \begin{Vmatrix} 1 & i & 0 \\ -i & 1 & 0 \\ 0 & 0 & 0 \end{Vmatrix} + \kappa_1 \begin{Vmatrix} 0 & 0 & 1 \\ 0 & 0 & -i \\ 1 & i & 0 \end{Vmatrix},$$

$$\delta_1 = \frac{1}{2}\sqrt{\frac{\pi}{2}} \cdot \frac{\omega_{p+}^2}{\omega^2 \cos\theta} e^{-z^2 1+}, \qquad \kappa_1 = -\left(1 - \frac{\omega}{\omega_{c+}}\right)\delta_1 \tan\theta, \tag{91}$$

and the cold dielectric coefficients, the ion and electron thermal corrections in Equation (90) are defined in Equations (81), (13) and (87), respectively. Then the dispersion Equation (6) takes on the form, making the damping corrections to Equation (14),

$$(\epsilon_+ A_{1+} + \epsilon_- A_{1-})n^7 + (A_0 + \epsilon_+ B_{1+} + \epsilon_- B_{1-})n^5$$
$$+ (B_0 + \epsilon_+ C_{1+} + \epsilon_- C_{1-})n^3 + C_0 n$$
$$+ i\epsilon_+^{-1/2}\delta_1(A_{-1+}n^4 + B_{-1+}n^2 + C_{-1+}) = 0, \tag{92}$$

where

$$A_{-1+} = -\left(1 - \frac{2\omega}{\omega_{c+}}\right)\sin^2\theta,$$
$$-B_{-1+} = \varepsilon_{33}^{(0)}(1 + \cos^2\theta),$$
$$C_{-1+} = \varepsilon_{33}^{(0)}(\varepsilon_{11}^{(0)} + i\varepsilon_{12}^{(0)}), \tag{93}$$

and the coefficients A_0, B_0, C_0; A_{1+}, B_{1+}, C_{1+}; A_{1-}, B_{1-}, C_{1-} are defined in Equations (84), (15), and (89), respectively.

As in Equation (43), i.e. the solution of Equation (41) for electron cyclotron damping, we now obtain the formal solution of Equation (92) for ion cyclotron damping in the form

$$n_{o,e}^2 = n_{o,e}^{(0)2}\left(1 \mp \sum_{+,-} \epsilon_{\pm}\frac{A_{1\pm}n_{o,e}^{(0)4} + B_{1\pm}n_{o,e}^{(0)2} + C_{1\pm}}{\sqrt{B_0^2 - 4A_0 C_0}} \mp i\,\frac{\epsilon_+^{-1/2}\delta_1}{n_{o,e}^{(0)3}}\right.$$
$$\left. \cdot\frac{A_{-1+}n_{o,e}^{(0)4} + B_{-1+}n_{o,e}^{(0)2} + C_{-1+}}{\sqrt{B_0^2 - 4A_0 C_0}}\right), \tag{94}$$

where the second term in the bracket represents the ion and electron thermal corrections to the real part of the refractive index, while the imaginary part represents ion cyclotron damping.

Recalling the frequency range $\omega \lesssim \omega_{c+} \ll \sqrt{(\omega_{c+}\,\omega_{c-})}$ in this subsection and referring to Equations (81), the zeroth order refractive indices in Equation (85) may be written explicitly as

$$n_o^{(0)2} = \frac{n_A^2}{1 - \omega^2/\omega_{c+}^2}\cdot\frac{1 + \cos^2\theta}{\cos^2\theta}\left\{1 - \left(1 - \frac{\omega^2}{\omega_{c+}^2}\right)\frac{\cos^2\theta}{(1 + \cos^2\theta)^2}\right\},$$

$$n_e^{(0)2} = \frac{n_A^2}{1 + \cos^2\theta}, \tag{95}$$

where n_A is the Alfvén index defined in Equation (4). In the vicinity of the ion cyclotron frequency ($\omega \approx \omega_{c+}$), the cold refractive index of the pure (slow) Alfvén wave, Equation (95), may be rewritten as

$$n_o^{(0)2} \simeq \frac{n_A^2}{1 - \omega/\omega_{c+}} \cdot \frac{1 + \cos^2 \theta}{2 \cos^2 \theta}. \tag{96}$$

With particular reference to 'ion cyclotron waves' or 'micropulsations', we again set $\omega \lesssim \omega_{c+}$, $\theta \simeq 0$ in Equations (94) and (96) and obtain a simpler expression for the pure (slow) Alfvén wave

$$n_o \simeq \frac{\omega_{p+}}{\sqrt{\omega_{c+}(\omega_{c+} - \omega)}}$$

$$\cdot \left\{ 1 + \frac{\epsilon_+}{2} \cdot \frac{\omega_{p+}^2 \omega}{(\omega_{c+} - \omega)^3} + \frac{i}{2}\sqrt{\frac{\pi}{2\epsilon_+}} \cdot \frac{\sqrt{(\omega_{c+} - \omega)^3}}{\omega_{p+}\sqrt{\omega}} e^{-z^2_{1+}} \right\}. \tag{97}$$

This formula applies under the condition $\sqrt{\eta} \gg |1 - \omega_{c+}/\omega| \gg \eta$; otherwise, the electron thermal corrections also become appreciable.

5.2. Ion cyclotron damping: ion cyclotron resonance

A treatment similar to that of the electron cyclotron resonances in Subsection 4.4 applies to ion cyclotron resonances. When the wave frequency is close enough to the ion cyclotron frequency, one of the quasi-cold conditions no longer holds, i.e. $|z_{1+}| \lesssim 1$. Then the treatment in the preceding subsection fails and expressions for the dielectric coefficients are altered radically as follows

$$\tilde{\varepsilon} = \tilde{\varepsilon}^{(0)} + i \frac{\tilde{\gamma}_+}{n\sqrt{\epsilon_+}}, \tag{98}$$

where

$$\hat{\varepsilon}_{11}^{(0)} \simeq \hat{\varepsilon}_{22}^{(0)} \simeq 1 - \frac{\omega_{p+}^2}{2\omega(\omega + \omega_{c+})},$$

$$\hat{\varepsilon}_{33}^{(0)} \simeq \varepsilon_{33}^{(0)} = 1 - \frac{\omega_{p+}^2 + \omega_{p-}^2}{\omega^2} \simeq - \frac{\omega_{p-}^2}{\omega^2},$$

$$\hat{\varepsilon}_{12}^{(0)} = - \hat{\varepsilon}_{21}^{(0)} \simeq - \frac{i3\omega_{p+}^2}{2\omega(\omega + \omega_{c+})},$$

$$\hat{\varepsilon}_{13}^{(0)} = \hat{\varepsilon}_{31}^{(0)} \simeq \frac{\omega_{p+}^2}{2\omega_{c+}\omega} \tan \theta,$$

$$\hat{\varepsilon}_{23}^{(0)} = - \hat{\varepsilon}_{32}^{(0)} \simeq - \frac{i\omega_{p+}^2}{2\omega_{c+}\omega} \tan \theta \tag{99}$$

and

$$\tilde{\gamma}_+ = \delta_1 \begin{Vmatrix} 1 & i & 0 \\ -i & 1 & 0 \\ 0 & 0 & 0 \end{Vmatrix}, \quad \delta_1 = \frac{1}{2}\sqrt{\frac{\pi}{2}} \frac{\omega_{p+}^2}{\omega^2 \cos \theta} e^{-z^2_{1+}}. \tag{100}$$

Then the dispersion relation (6) takes on the form

$$A_0 n^5 + B_0 n^3 + C_0 n + i\epsilon_+^{-1/2}\delta_1 \left(A_{-1+}n^4 + B_{-1+}n^2 + C_{-1+}\right) = 0, \quad (101)$$

where

$$A_{-1+} = \sin^2\theta,$$
$$-B_{-1+} = \varepsilon_{33}^{(0)}(1 + \cos^2\theta) + 2(\hat{\varepsilon}_{11}^{(0)} + i\hat{\varepsilon}_{12}^{(0)})\sin^2\theta,$$
$$C_{-1+} = 2\varepsilon_{33}^{(0)}(\hat{\varepsilon}_{11}^{(0)} + i\hat{\varepsilon}_{12}^{(0)}), \quad (102)$$

and A_0, B_0, C_0 are defined by setting $\hat{\varepsilon}_{ik}^{(0)}$ instead of ε_{ik} in Equations (7).

Retaining only terms proportional to $\epsilon_+^{-1/2}$, the solution of Equation (101) may be written as

$$n_{o,e}^{(0)2} = 1 - \frac{2(A_{-1+} + B_{-1+} + C_{-1+})}{2A_{-1+} + B_{-1+} \pm \sqrt{B_{-1+}^2 - 4A_{-1+}C_{-1+}}}. \quad (103)$$

In the next approximation we find

$$n_{o,e}^2 = n_{o,e}^{(0)2}\left(1 \pm i\epsilon_+^{1/2}\frac{A_0 n_{o,e}^{(0)4} + B_0 n_{o,e}^{(0)2} + C_0}{\delta_1 n_{o,e}^{(0)}\sqrt{B_{-1+}^2 - 4A_{-1+}C_{-1+}}}\right)$$
$$\text{for} \quad \epsilon_+^{-1/2}|\delta_1|\sin^2\theta \gg 1. \quad (104)$$

Thus the pure and modified Alfvén waves are damped in the immediate vicinity of the ion cyclotron frequency for oblique propagation and the magnitude of the damping coefficient is of order $\epsilon_+^{1/2}$.

5.3. ELECTRON LANDAU DAMPING

In Subsection 4.5 electron Landau damping was introduced for the electron plasma waves and slow electromagnetic waves in the HF and VLF ranges and was seen to be significant when their phase velocity approaches the electron thermal velocity. Similar considerations apply to the ELF range. As was indicated in Section 2, in fact, the Alfvén speed is comparable with the electron thermal velocity at heights of $300 \sim 500$ km and about 20000 km. This indicates that electron Landau damping may occur for both pure and modified Alfvén waves in the ELF range around these heights.

We consider now electron Landau damping for the pure and modified Alfvén waves in the case when $|z_{0-}| \gtrsim 1$. Then, among the elements of the dielectric tensor, the damping correction is required only for the dielectric coefficient ε_{33} as defined in Equation (52). Inclusion of the electron Landau damping term in Equation (6) yields, referring to Equations (7), (81), and (52),

$$A_0 n^7 + B_0 n^5 + C_0 n^3 + i\varepsilon_-^{-3/2}\delta_{0-}\left(A_{0-}n^4 + B_{0-}n^2 + C_{0-}\right) = 0, \quad (105)$$

where

$$A_{0-} = \cos^2\theta, \quad B_{0-} = \varepsilon_{11}^{(0)}(1 + \cos^2\theta), \quad C_{0-} = \varepsilon_{11}^{(0)2} + \varepsilon_{12}^{(0)2},$$

$$\delta_{0-} = \sqrt{\frac{\pi}{2}} \cdot \frac{\omega_{p-}^2}{\omega^2} \cdot \frac{e^{-z_{0-}^2}}{\cos^3\theta} \quad (106)$$

and A_0, B_0, C_0 are defined in Equations (84).

Since the solutions of Equation (105) for $\delta_{0-} = 0$ form the refractive indices of the pure and modified Alfvén waves for zero electron Landau damping, the damping correction for Equation (85) leads to

$$n_{o,e}^2 = n_{o,e}^{(0)2}\left\{1 \mp i\,\frac{\epsilon_-^{-3/2}\delta_{0-}\left(A_{0-}n_{o,e}^{(0)4} + B_{0-}n_{o,e}^{(0)2} + C_{0-}\right)}{n_{o,e}^{(0)5}\sqrt{B_0^2 - 4A_0C_0}}\right\}, \tag{107}$$

where $n_{o,e}^{(0)}$ is the index for zero Landau damping defined in Equation (85).

5.4. ELECTRON LANDAU AND WEAK ION CYCLOTRON DAMPING

For the case when electron Landau damping predominates ($|z_{0-}| \gtrsim 1$) near but not too near the ion cyclotron frequency ($|z_{1+}| \gtrsim 1$), Equation (6) takes on the form, referring to Equations (92) and (105) and taking into account electron Landau and ion cyclotron damping,

$$A_0 n^7 + B_0 n^5 + C_0 n^3 + i\{\epsilon_-^{3/2}\delta_{0-}\left(A_{0-}n^4 + B_{0-}n^2 + C_{0-}\right)$$
$$+ \epsilon_+^{-1/2}\delta_1 n^2\left(A_{-1+}n^4 + B_{-1+}n^2 + C_{-1+}\right)\} = 0, \tag{108}$$

where the terms with δ_{0-} and δ_1 refer to electron Landau and ion cyclotron damping, respectively, and the thermal corrections to the real parts of the indices were all neglected. Since the solutions of Equation (108) for $\delta_{0-}=0$ or $\delta_1=0$ are given by Equation (94) (set the real thermal corrections equal to zero) or (107), respectively, both damping corrections yield

$$n_{o,e}^2 = n_{o,e}^{(0)2}\left\{1 \mp i\,\frac{\epsilon_-^{3/2}\delta_{0-}\left(A_{0-}n_{o,e}^{(0)4} + B_{0-}n_{o,e}^{(0)2} + C_{0-}\right)}{n_{o,|:}^{(0)5}\sqrt{B_0^2 - 4A_0C_0}}\right.$$
$$\left.\mp i\,\frac{\epsilon_+^{-1/2}\delta_1\left(A_{-1+}n_{o,e}^{(0)4} + B_{-1+}n_{o,e}^{(0)2} + C_{-1+}\right)}{n_{o,e}^{(0)3}\sqrt{B_0^2 - 4A_0C_0}}\right\}. \tag{109}$$

The second and the third term represent spatial electron Landau and ion cyclotron damping, respectively, and $n_{o,e}^{(0)}$ is defined in Equations (95) and (96).

5.5. SECOND HARMONIC ION CYCLOTRON RESONANCE

When the wave frequency falls close enough to the second harmonic of the ion cyclotron frequency, one of the quasi-cold conditions no longer holds, i.e. $|z_{2+}| \ll 1$. Then, expressions for the dielectric coefficients (10), (11), (13), (81) are altered and the same procedure as carried out for the second harmonic electron cyclotron resonance in Subsection 4.8 yields

$$\tilde{\varepsilon} = \tilde{\tilde{\varepsilon}} + in\sqrt{\epsilon_+}\tilde{\gamma}_+, \qquad \tilde{\tilde{\varepsilon}} = \tilde{\varepsilon}^{(0)} + \tilde{\varepsilon}^{(2c)}, \tag{110}$$

where

$$\tilde{\varepsilon}^{(2c)} = \kappa_2 \begin{Vmatrix} 1 & -i & 0 \\ i & 1 & 0 \\ 0 & 0 & 0 \end{Vmatrix}, \qquad \tilde{\gamma}_+ = \delta_2 \begin{Vmatrix} 1 & -i & 0 \\ i & 1 & 0 \\ 0 & 0 & 0 \end{Vmatrix},$$

$$\kappa_2 = -\frac{\omega_{p+}^2(\omega - 2\omega_{c+})}{2\omega_{c+}^2\omega}\tan^2\theta, \qquad \delta_2 = \frac{1}{2}\sqrt{\frac{\pi}{2}}\cdot\frac{\omega_{p+}^2}{\omega_{c+}^2}\cdot\frac{\sin^2\theta}{\cos\theta}e^{-z^2{}_{2+}} \tag{111}$$

and $\tilde{\varepsilon}^{(0)}$ represents the cold dielectric tensor defined in Equations (81).

Then, the dispersion Equation (6) takes on the form

$$A_0 n^4 + B_0 n^2 + C_0 + i\epsilon_+^{-1/2}\delta_2 n (A_{-2+}n^4 + B_{-2+}n^2 + C_{-2+}) = 0, \quad (112)$$

where the coefficients A_0, B_0, C_0, $A_{-2+} = A_{-2-}$, $B_{-2+} = B_{-2-}$, $C_{-2+} = C_{-2-}$ are defined in Equations (17) and (67) but the zeroth order coefficients $\varepsilon_{ik}^{(0)}$ must be replaced by $\hat{\varepsilon}_{ik}$ defined in Equation (110). Then, the solution of Equation (112) may be written as

$$n_{o,e}^2 = n_{o,e}^{(0)2}\left\{1 \mp i\epsilon_+^{1/2}\frac{\delta_2(A_{-2+}n_{o,e}^{(0)4} + B_{-2+}n_{o,e}^{(0)2} + C_{-2+})}{n_{o,e}^{(0)}\sqrt{B_0^2 - 4A_0C_0}}\right\}, \quad (113)$$

where $n_{o,e}^{(0)2}$ is defined in Equation (85) or (95). Thus the Alfvén waves are damped at the second harmonic of the ion cyclotron frequency. The magnitude of damping is of order $\epsilon_+^{1/2}$ and is equal to the order of cyclotron damping at the ion cyclotron frequency for oblique propagation.

5.6. LOWER HYBRID RESONANCES

In Subsection 4.10 the hybrid resonances were defined for perpendicular propagation for which there is no Landau and cyclotron damping. We consider now the lower hybrid resonances for quasi-perpendicular propagation in the ELF range.

Generally, for oblique propagation the hybrid resonances are found by setting

$$A_0 = \varepsilon_{11}^{(0)}\sin^2\theta + \varepsilon_{33}^{(0)}\cos^2\theta = 0, \quad (114)$$

where $\varepsilon_{11}^{(0)}$, $\varepsilon_{33}^{(0)}$ are defined in Equations (11). Then Equation (114) may be written as

$$\tan^2\theta = -\frac{(\omega^2 - \omega_p^2)(\omega^2 - \omega_{c+}^2)(\omega^2 - \omega_{c+}^2)}{\omega^2(\omega^2 - \omega_h^2)(\omega^2 - \omega_H^2)}, \quad (115)$$

where $\omega_p^2 = \omega_{p+}^2 + \omega_{p-}^2 \simeq \omega_{p-}^2$, and ω_h, ω_H are the lower and upper hybrid frequencies defined in Equations (79).

For the ELF range where $\omega \ll \omega_{c-}$, ω_{p-}, the solutions of Equation (115) define the lower hybrid frequencies, i.e.

$$\omega_\pm^2 = \frac{1}{2\omega_H^2\sin^2\theta}\{\omega_p^2\omega_{c-}^2\cos^2\theta + \omega_h^2\omega_H^2\sin^2\theta$$
$$\pm \sqrt{(\omega_p^2\omega_{c-}^2\cos^2\theta + \omega_h^2\omega_H^2\sin^2\theta)^2 - \omega_p^2\omega_{c+}^2\omega_{c-}^2\omega_H^2\sin^2 2\theta}\}. \quad (116)$$

Generally the lower hybrid resonances for oblique propagation are difficult to observe because of large Landau damping at intermediate angles.

For quasi-perpendicular propagation, i.e. when the angle θ is close to $\pi/2$, however, Landau damping is so small as to allow observation of these resonances. Then the lower hybrid frequencies may be rewritten, with $\cos\theta \ll 1$ in Equation (116), as

$$\omega_\pm^2 = \frac{\omega_{c-}^2}{2(\omega_{p-}^2 + \omega_{c-}^2)}\{\omega_{p-}^2\cos^2\theta + \omega_{p+}^2 + \omega_{c+}^2$$
$$\pm \sqrt{(\omega_{p-}^2\cos^2\theta + \omega_{p+}^2 + \omega_{c+}^2)^2 - 4\omega_{p+}^2(\omega_{p+}^2 + \omega_{c+}\omega_{c-})\cos^2\theta}\}. \quad (117)$$

Further, if $\cos^2 \theta \ll \eta$, Equation (117) may be simplified as

$$\omega_+ = \omega_{c-} \sqrt{\frac{\omega_{p+}^2 + \omega_{c+}^2}{\omega_{p+}^2 + \omega_{p-}^2 + \omega_{c-}^2}} \simeq \sqrt{\frac{\omega_{c+}\omega_{c-}}{1 + \omega_{c-}^2/\omega_{p-}^2}} \simeq \sqrt{\omega_{c+}\omega_{c-}},$$

$$\omega_- = \frac{\omega_{p-}\omega_{c+}}{\sqrt{\omega_{p+}^2 + \omega_{c+}^2}} \cos\theta \simeq \sqrt{\omega_{c+}\omega_{c-}} \cos\theta \simeq \omega_+ \cos\theta \ll \omega_{c+}, \qquad (118)$$

where ω_+ represents the upper branch of the lower hybrid resonances and is identical with the lower hybrid resonance for perpendicular propagation, while ω_- corresponds to the lower branch of the lower hybrid resonances which can be observed because of small Landau damping for quasi-perpendicular propagation. Further, the upper branch is associated with the extraordinary or modified (fast) Alfvén wave, and the lower with the ordinary or pure (slow) Alfvén wave. The lower branch has a band frequency spectrum well below the ion cyclotron frequency rather than a sharp line spectrum, but its frequency spread is limited by Landau damping.

In the vicinity of the upper branch of the lower hybrid resonances but not too close to ω_+, the refractive index of the extraordinary or the modified Alfvén wave takes on the form, referring to Equations (107), (81), and (84),

$$n_e^2 = n_e^{(0)2} \left\{ 1 + i \frac{\epsilon_-^{-3/2}\delta_{0-}(A_{0-}n_e^{(0)4} + B_{0-}n_e^{(0)2} + C_{0-})}{n_e^{(0)5}B_0} \right\}, \qquad (119)$$

where

$$n_e^{(0)2} \simeq -\frac{B_0}{A_0 + \epsilon_+ B_{1+} + \epsilon_- B_{1-}}$$

$$\simeq \frac{\omega_p^4 \omega^2}{\omega^2 \{\omega_p^2 \omega_{c-}(\omega_{c+} + \omega_{c-}\cos^2\theta) - \omega^2(\omega_p^2 + \omega_{c-}^2)\} + \epsilon_- \xi \omega_p^4 \omega_{c+}^2}$$

$$(\omega_p^2 = \omega_{p+}^2 + \omega_{p-}^2) \qquad (120)$$

and the imaginary part expresses the effect of electron Landau damping. In this frequency range, the refractive index of the plasma wave leads, referring to Equations (23), (55), (13), and (15), to

$$n_p^2 = n_p^{(0)2} \left\{ 1 + i \frac{\epsilon_-^{-3/2}\delta_{0-}}{n_p^{(0)}(A_0 n_p^{(0)2} + 2B_0)} \right\}, \qquad (121)$$

where

$$n_p^{(0)2} = -\frac{A_0}{\epsilon_+ A_{1+} + \epsilon_- A_{1-}}$$

$$\simeq \frac{\omega_{c-}^2 \{\omega_p^2 \omega_-(\omega_{c+} + \omega_{c-}\cos^2\theta) - \omega^2(\omega_p^2 + \omega_{c-}^2)\}}{\epsilon_- \omega_p^2(3\omega^4 - \xi\omega_{c+}^2\omega_{c-}^2)} \qquad (122)$$

and the imaginary part also represents electron Landau damping. Equations (119) to (122) indicate that the fast Alfvén wave tends to couple to the plasma wave, and in the immediate vicinity of the lower hybrid frequency ω_+, the refractive index of the coupled waves may be written as

$$n_{e,p}^2 = \pm \sqrt{-\frac{B_0}{\epsilon_+ A_{1+} + \epsilon_- A_{1-}}} = \pm \frac{i\omega_p \omega_{c-}}{\sqrt{\epsilon_-(3\omega^4 - \xi\omega_{c+}^2\omega_{c-}^2)}}. \qquad (123)$$

This indicates that the fast Alfvén and plasma waves are strongly damped close enough to the frequency ω_+.

In the vicinity of the lower branch of the lower hybrid resonances, $\omega_- = \sqrt{\omega_{c+}\,\omega_{c-}}$ $\cos\theta \ll \omega_{c+}$, the refractive indices of the slow and fast Alfvén waves may be written, referring to Equations (85) and (107), as

$$n_{o,e}^2 = n_{o,e}^{(0)2}\left\{1 \mp i\,\frac{\epsilon_-^{3/2}\delta_{0-}(A_{0-}n_{o,e}^{(0)4} + B_{0-}n_{o,e}^{(0)2} + C_{0-})}{n_{o,e}^{(0)5}B_0}\right\}, \qquad (124)$$

where for $\omega \ll \omega_{c+}$, setting $\varepsilon_{12}^{(0)} \simeq 0$ in Equation (85)

$$n_o^{(0)} \simeq n_A/\cos\theta, \qquad n_e^{(0)} \simeq n_A. \qquad (125)$$

In this frequency range the refractive index of the plasma wave takes on the form, referring to Equations (23), (55), (13), and (15),

$$n_p^2 = n_p^{(0)2}\left\{1 + i\,\frac{\epsilon_-^{-3/2}\delta_{0-}}{n_p^{(0)}(A_0 n_p^{(0)2} + 2B_0)}\right\}, \qquad (126)$$

where

$$n_p^{(0)2} = -\frac{A_0}{\epsilon_+ A_{1+} + \epsilon_- A_{1-}} = \frac{\omega_{c+}^2\{\omega_p^2\omega_{c+}^2\cos^2\theta - \omega^2(\omega_{p+}^2 + \omega_{c+}^2)\}}{3\epsilon_+\omega_{p+}^2\omega^4(1 + \eta^2/\xi)}. \qquad (127)$$

The imaginary part in Equation (124) or (126) again represents electron Landau damping. Equations (124) to (127) indicate that the slow Alfvén wave tends to couple to the plasma wave.

5.7. MAGNETOSONIC REGIME (I)

So far we have treated waves whose phase velocity is faster than or at least comparable to the electron thermal velocity ($|z_{0-}| \gg 1$ or $|z_{0-}| \gtrsim 1$). We come now to a region of the magnetosonic regime in which the wave frequency is well below the ion cyclotron frequency ($\omega \ll \omega_{c+}$) and the wave phase velocity small compared to the electron thermal velocity, although still faster than the ion thermal velocity (i.e. $|z_{0-1}| \ll 1$, but still $|z_{0+}| \gg 1$ and $|z_{p\pm}| \gg 1$ ($p \neq 0$) together with $|\mu_\pm| \ll 1$). The pure (slow) Alfvén wave is still given by the first equation of (125) but the modified (fast) Alfvén wave is altered. Basically, expressions for the dielectric coefficients (10), (11), (13), (52), and (81) are altered and the use of the convergent series expansion for the complex error functions concerned with the parameter z_{0-} yields, for the dielectric coefficients ε_{22}, ε_{33}, ε_{23}, and ε_{32},

$$\varepsilon_{22} = \hat{\varepsilon}_{22}^{(0)} + in\sqrt{\epsilon_+}\,\delta,$$

$$\varepsilon_{33} = \hat{\varepsilon}_{33}^{(0)} + \frac{\gamma}{\in_+ n^2},$$

$$\varepsilon_{23} = -\varepsilon_{32} = \hat{\varepsilon}_{23}^{(0)}, \tag{128}$$

where

$$\hat{\varepsilon}_{22}^{(0)} = 1 + \frac{\omega_{p+}^2}{\omega_{c+}^2}(1 - 2\eta \tan^2 \theta) \simeq 1 + \frac{\omega_{p+}^2}{\omega_{c+}^2} = \varepsilon_{11}^{(0)} \quad (\text{for} \quad \cos^2 \theta \gg \eta),$$

$$\hat{\varepsilon}_{33}^{(0)} = 1 - \frac{\omega_{p+}^2}{\omega^2},$$

$$\hat{\varepsilon}_{23}^{(0)} = -i \frac{\omega_{p+}^2}{\omega_{c+}\omega} \tan \theta \tag{129}$$

and

$$\gamma = \frac{\xi \omega_{p+}^2}{\omega^2 \cos^2 \theta}, \qquad \delta = \sqrt{\frac{2\pi\eta}{\xi}} \cdot \frac{\omega_{p+}^2}{\omega_{c+}^2} \cdot \frac{\sin^2 \theta}{\cos \theta}. \tag{130}$$

The thermal corrections arising from parameters other than z_{0-} were all neglected. Substituting Equations (128) into Equations (7) and making use of the inequalities

$$|\varepsilon_{12}^{(0)}/\varepsilon_{11}^{(0)}| \ll 1, \qquad |\varepsilon_{11}^{(0)}/\varepsilon_{33}^{(0)}| \ll 1, \tag{131}$$

the dispersion relation (6) takes on the form

$$\in_+ A_0 n^6 + (A_1 + \in_+ B_0)n^4 + (B_1 + \in_+ C_0)n^2 + C_1$$
$$+ in\sqrt{\in_+}\delta(A_{-1}n^4 + B_{-1}n^2 + C_{-1}) = 0, \tag{132}$$

where

$$A_0 = \varepsilon_{11}^{(0)}\sin^2\theta + \hat{\varepsilon}_{33}^{(0)}\cos^2\theta \simeq \hat{\varepsilon}_{33}^{(0)}\cos^2\theta,$$

$$-B_0 = \varepsilon_{11}^{(0)}(\hat{\varepsilon}_{33}^{(0)} + \hat{\varepsilon}_{22}^{(0)}\sin^2\theta) + (\hat{\varepsilon}_{22}^{(0)}\hat{\varepsilon}_{33}^{(0)} + \hat{\varepsilon}_{23}^{(0)2})\cos^2\theta$$
$$\simeq \varepsilon_{11}^{(0)}\hat{\varepsilon}_{33}^{(0)}(1 + \cos^2\theta) + \hat{\varepsilon}_{23}^{(0)2}\cos^2\theta,$$

$$C_0 = \varepsilon_{11}^{(0)}(\hat{\varepsilon}_{22}^{(0)}\hat{\varepsilon}_{33}^{(0)} + \hat{\varepsilon}_{23}^{(0)2}) \simeq \varepsilon_{11}^{(0)}(\varepsilon_{11}^{(0)}\hat{\varepsilon}_{33}^{(0)} + \hat{\varepsilon}_{23}^{(0)2}), \tag{133}$$

$$A_1 = \gamma \cos^2\theta,$$

$$-B_1 = \gamma(\varepsilon_{11}^{(0)} + \hat{\varepsilon}_{22}^{(0)}\cos^2\theta) \simeq \gamma\varepsilon_{11}^{(0)}(1 + \cos^2\theta),$$

$$C_1 = \gamma\varepsilon_{11}^{(0)}\hat{\varepsilon}_{22}^{(0)} \simeq \gamma\varepsilon_{11}^{(0)2}, \tag{134}$$

and

$$-A_{-1} = \varepsilon_{11}^{(0)}\sin^2\theta + \hat{\varepsilon}_{33}^{(0)}\cos^2\theta \simeq \hat{\varepsilon}_{33}^{(0)}\cos^2\theta,$$

$$B_{-1} = \varepsilon_{11}^{(0)}\hat{\varepsilon}_{33}^{(0)} - \gamma\cos^2\theta,$$

$$C_{-1} = \gamma\varepsilon_{11}^{(0)}. \tag{135}$$

The above approximations hold when $\cos^2\theta \gg \eta$.

Then the zeroth order solution of Equation (132) for $\in_+ = 0$ and $\delta = 0$ may be written as

$$n_{o,e}^{(0)2} = \frac{1}{2A_1}(-B_1 \pm \sqrt{B_1^2 - 4A_1C_1})$$

$$= 1 - \frac{2(A_1 + B_1 + C_1)}{2A_1 + B_1 \pm \sqrt{B_1^2 - 4A_1 C_1}} . \tag{136}$$

Substituting Equations (134) into Equations (136) we obtain the zeroth order refractive indices of the pure and modified Alfvén waves in the form

$$n_o^{(0)2} = \varepsilon_{11}/\cos^2 \theta \simeq (1 + n_A^2)/\cos^2 \theta \simeq n_A^2/\cos^2 \theta ,$$
$$n_e^{(0)2} = \hat{\varepsilon}_{22}^{(0)} = 1 + n_A^2 (1 - 2\eta \tan^2 \theta) . \tag{137}$$

Comparing these equations with Equations (125), one sees that the pure Alfvén wave is still given by the same formula, though the modified Alfvén wave is somewhat altered.

Since the Alfvén speed v_A is small compared to the electron thermal velocity in this magnetosonic regime, the modified Alfvén wave tends to couple with the ion plasma wave. Then the refractive indices of the modified Alfvén and ion plasma waves may be written, retaining the first three terms with respect to n^2 in Equation (132), as

$$n_{e,p}^{(0)2} = \frac{1}{2\epsilon_+ A_0}$$
$$\cdot \{ - (A_1 + \epsilon_+ B_0) \pm \sqrt{(A_1 + \epsilon_+ B_0)^2 - 4\epsilon_+ A_0 (B_1 + \epsilon_+ C_0)} \} . \tag{138}$$

Making use of Equations (133) and (134), Equation (138) takes on the explicit form

$$n_{e,p}^{(0)2} = \frac{2\epsilon_+ n_A^2 + \xi(1 \mp \sqrt{1 - 4\epsilon_+/\xi \cdot n_A^2 \cos^2 \theta})}{2\epsilon_+ \cos^2 \theta} . \tag{139}$$

The modified Alfvén and ion plasma waves defined in this equation are often referred to as the fast and slow magnetosonic waves.

Taking now into account the electron Landau damping terms in Equation (132), the damping corrections to Equations (137) and (139) yield

$$n_{o,e}^2 = n_{o,e}^{(0)2} \left\{ 1 \mp i \frac{\sqrt{\epsilon_+} \delta(A_{-1} n_{o,e}^{(0)4} + B_{-1} n_{o,e}^{(0)2} + C_{-1})}{n_{o,e}^{(0)} \sqrt{B_1^2 - 4A_1 C_1}} \right\} , \tag{140}$$

$$n_{e,p}^2 = n_{e,p}^{(0)2} \left\{ 1 - i \frac{\sqrt{\epsilon_+} \delta(A_{-1} n_{e,p}^{(0)4} + B_{-1} n_{e,p}^{(0)2} + C_{-1})}{n_{e,p}^{(0)3} (2\epsilon_+ A_0 n_{e,p}^{(0)2} + A_1 + \epsilon_+ B_0)} \right\} , \tag{141}$$

where $n_{o,e}^{(0)2}$, $n_{e,p}^{(0)2}$ are defined in Equations (137) and (139), respectively.

5.8. MAGNETOSONIC REGIME (II)

In the second class of the magnetosonic regime, the wave phase velocity is small compared to the thermal velocities of both ions and electrons, i.e. $|z_{0+}| \ll 1$, $|z_{0-}| \ll 1$, but other quasi-cold conditions hold, i.e. $|z_{p\pm}| \gg 1$ ($p \neq 0$), $|\mu_\pm| \ll 1$. The wave frequency is assumed to be well below the ion cyclotron frequency ($\omega \ll \omega_{c+}$). The pure Alfvén wave is still described by the first equation of (125) but the modified Alfvén wave is again altered, since the convergent series expansion for the complex error functions concerned with the parameters z_{0+} and z_{0-} must now be used to evaluate the dielectric coefficients. Then new dielectric coefficients ε_{22}, ε_{33} may be written as

$$\varepsilon_{22} = \hat{\varepsilon}_{22}^{(0)} + in\sqrt{\epsilon_+}\,\delta,$$

$$\varepsilon_{33} = 1 + \frac{\gamma}{\epsilon_+ n^2}, \tag{142}$$

where

$$\hat{\varepsilon}_{22}^{(0)} = 1 + \frac{\omega_{p+}^2}{\omega_{c+}^2}\{1 - 2(1+\eta)\tan^2\theta\},$$

$$\gamma = \frac{\omega_{p+}^2(1+\xi)}{\omega^2\cos^2\theta},$$

$$\delta = \sqrt{2\pi}\,\frac{\omega_{p+}^2}{\omega_{c+}^2}\left(1 + \sqrt{\frac{\eta}{\xi}}\right)\frac{\sin^2\theta}{\cos\theta}. \tag{143}$$

Substituting Equations (142) into Equation (7) and setting $\varepsilon_{12} = -\varepsilon_{21} \simeq 0$, $\varepsilon_{13} = \varepsilon_{31} \simeq 0$, $\varepsilon_{23} = -\varepsilon_{32} \simeq 0$, the dispersion relation (6) takes on the form

$$\epsilon_+ A_0 n^6 + (A_1 + \epsilon_+ B_0)n^4 + (B_1 + \epsilon_+ C_0)n^2 + C_1$$
$$+ in\sqrt{\epsilon_+}\,\delta(A_{-1}n^4 + B_{-1}n^2 + C_{-1}) = 0, \tag{144}$$

where

$$A_0 = \varepsilon_{11}^{(0)}\sin^2\theta + \cos^2\theta = 1 + n_A^2\sin^2\theta,$$
$$-B_0 = \varepsilon_{11}^{(0)}(1 + \hat{\varepsilon}_{22}^{(0)}\sin^2\theta) + \hat{\varepsilon}_{22}^{(0)}\cos^2\theta = 1 + \hat{\varepsilon}_{22}^{(0)} + n_A^2(1 + \hat{\varepsilon}_{22}^{(0)}\sin^2\theta),$$
$$C_0 = \varepsilon_{11}^{(0)}\hat{\varepsilon}_{22}^{(0)}, \tag{145}$$

$$A_1 = \gamma\cos^2\theta,$$
$$-B_1 = \gamma(\varepsilon_{11}^{(0)} + \hat{\varepsilon}_{22}^{(0)}\cos^2\theta),$$
$$C_1 = \gamma\varepsilon_{11}^{(0)}\hat{\varepsilon}_{22}^{(0)}, \tag{146}$$

and

$$-A_{-1} = \varepsilon_{11}^{(0)}\sin^2\theta + \cos^2\theta = 1 + n_A^2\sin^2\theta,$$
$$B_{-1} = \varepsilon_{11}^{(0)} - \gamma\cos^2\theta,$$
$$C_{-1} = \gamma\varepsilon_{11}^{(0)}. \tag{147}$$

Substituting Equations (146) into Equation (144) for $\epsilon_+ = 0$ and $\delta = 0$, the zeroth-order refractive indices of the pure and modified Alfvén waves lead to

$$n_o^{(0)2} = \frac{\varepsilon_{11}}{\cos^2\theta} \simeq \frac{1 + n_A^2}{\cos^2\theta} \simeq \frac{n_A^2}{\cos^2\theta},$$
$$n_e^{(0)2} = \hat{\varepsilon}_{22}^{(0)} = 1 + n_A^2\{1 - 2(1+\eta)\tan^2\theta\}. \tag{148}$$

One sees that the slow Alfvén wave is still described by the same expression as the first equation of (137) but the fast Alfvén wave is further altered.

By a similar treatment as in the preceding subsection, one can obtain the refractive indices of the fast and slow magnetosonic waves and the corrections for ion and electron Landau damping in the forms (138) and (141), where the coefficients are defined in Equations (144) to (147).

5.9. ION PLASMA WAVES

As was indicated, for instance in Equation (21) or (139), the longitudinal plasma waves in a magnetic field cannot be strictly separated from transverse electromagnetic waves under the quasi-cold conditions. In the limiting case $|n| \gg 1$, however, we can distinguish longitudinal plasma waves whose dispersion equation may be written approximately, referring to Equations (6) and (7), as

$$A = \varepsilon_{11} \sin^2 \theta + \varepsilon_{33} \cos^2 \theta + \varepsilon_{13} \sin 2\theta = 0 \,. \tag{149}$$

We go now to the ion plasma waves propagating at an oblique angle in a magnetic field within the magnetosonic regime (I) ($|z_{0-}| \ll 1$, $|z_{0+}| \gg 1$, $|z_{p\pm}| \gg 1$ ($p \neq 0$), particularly $|z_{1+}| \gg 1$). The wave frequency is not restricted to $\omega \ll \omega_{c+}$ but may be $\omega \lesssim \omega_{c+}$ or even $\omega_{c+} \lesssim \omega \ll \sqrt{(\omega_{c+} \omega_{c-})}$. Making use of Equations (81), and (128) to (130), Equation (149) for the ion plasma waves takes the form

$$\frac{1}{\omega^2} + \frac{\tan^2 \theta}{\omega^2 - \omega_{c+}^2} = \frac{1}{\cos^2 \theta} \left(\frac{1}{\omega_{p+}^2} + \frac{\xi}{\epsilon_+ n^2 \omega^2} \right) \,. \tag{150}$$

This determines two frequencies for the fast and slow ion plasma waves,

$$\omega_{1,2}^2 = \frac{1}{2} \{ (\omega_{10}^2 + \omega_{c+}^2) \pm \sqrt{(\omega_{10}^2 + \omega_{c+}^2)^2 - 4\omega_{10}^2 \omega_{c+}^2 \cos^2 \theta} \} \,, \tag{151}$$

where

$$\frac{1}{\omega_{10}^2} = \frac{1}{\omega_{p+}^2} + \frac{\xi}{\epsilon_+ n^2 \omega^2} = \frac{1}{\omega_{p+}^2} + \frac{1}{k^2 v_s^2} \quad \left(v_s = \sqrt{\frac{KT_-}{m_+}} \right) \tag{152}$$

and v_s is the isothermal sound velocity. Equation (152) is the dispersion equation of the ion plasma waves in the absence of a magnetic field and at the same time expresses the fast ion plasma wave propagating parallel to a magnetic field ($\theta = 0$).

From Equation (151) the frequency range of the fast ion plasma wave may be expressed in the limits of the angle from $\theta = 0$ to $\theta = \pi/2$

$$\omega_{10} = \frac{\omega_{p+}}{\sqrt{1 + \dfrac{\omega_{p-}^2}{\epsilon_- n^2 \omega^2}}} = \frac{\omega_{p+}}{\sqrt{1 + \dfrac{1}{k^2 \lambda_D^2}}}, \quad \omega_{1(\pi/2)} = \sqrt{\omega_{c+}^2 + \frac{\omega_{p+}^2}{1 + \dfrac{1}{k^2 \lambda_D^2}}}, \tag{153}$$

where

$$\lambda_D = \frac{v_s}{\omega_{p+}} = \frac{v_{T-}}{\omega_{p-}} = \sqrt{\frac{KT_- \varepsilon_0}{N_- Z^2 e^2}} \tag{154}$$

is the so-called Debye shielding distance.

For large wavelengths ($k\lambda_D \ll 1$), Equations (153) reduce to

$$\omega_{10} \simeq k v_s \gtrsim \omega_{c+}, \qquad \omega_{1(\pi/2)} \simeq \omega_{c+}, \tag{155}$$

while for small wavelengths ($k\lambda_D \gg 1$) Equations (153) reduce to

$$\omega_{10} \simeq \omega_{p+}, \qquad \omega_{1(\pi/2)} \simeq \sqrt{\omega_{p+}^2 + \omega_{c+}^2}. \tag{156}$$

Since $\omega_{p+} \gg \omega_{c+}$ (see Equation (1)), the frequency of the fast ion plasma wave spreads over a wide range from ω_{c+} to $\sqrt{(\omega_{p+}^2 + \omega_{c+}^2)}$ depending on the relative magnitude of the wavelength and the Debye shielding distance.

For the slow ion plasma wave from Equation (151) we have, when $k\lambda_D \gg 1$,

$$\omega_2 \simeq \omega_{c+} \cos\theta, \tag{157}$$

while when $k\lambda_D \ll 1$ and $k v_s \ll \omega_{c+}$ we have a slow magnetosonic wave

$$\omega_2 \simeq k v_s \cos\theta. \tag{158}$$

Going back to Equation (150), the zero-damped refractive indices of the fast and slow ion plasma waves may be written as

$$n_{1,2}^{(0)2} = \frac{\xi}{\epsilon_+} \cdot \frac{\omega_{p+}^2 (\omega_{1,2}^2 - \omega_{c+}^2)}{\omega_{p+}^2 (\omega_{1,2}^2 - \omega_{c+}^2 \cos^2\theta) - \omega_{1,2}^2 (\omega_{1,2}^2 - \omega_{c+}^2)}, \tag{159}$$

where $\omega_{1,2}$ is defined in Equation (151).

Since we are in the magnetosonic regime (I) ($|z_{0-}| \ll 1$, $|z_{0+}| \gg 1$) in the present subsection, ion Landau damping is small compared to electron Landau damping. We therefore take into account electron Landau damping only. Then the damping correction in Equation (149) occurs only in the dielectric coefficient ε_{33} as follows:

$$\varepsilon_{33} = 1 - \frac{\omega_{p+}^2}{\omega^2} + \frac{\xi \omega_{p+}^2}{\omega^2 \epsilon_+ n^2 \cos^2\theta} \left(1 + i \sqrt{\frac{\pi}{2}} \cdot \frac{1}{\sqrt{\epsilon_-}\, n \cos\theta} \right). \tag{160}$$

Thus the Landau damping correction to the refractive indices of the fast and slow ion plasma waves, Equation (159), yields

$$\begin{aligned}
n_{1,2}^2 &= n_{1,2}^{(0)2} \left(1 + i \sqrt{\frac{\pi}{2}} \cdot \frac{1}{\sqrt{\epsilon_-}\, n_{1,2}^{(0)} \cos\theta} \right) \\
&= n_{1,2}^{(0)2} \left[1 + i \frac{\sqrt{\eta}\, \{\omega_{p+}^2 (\omega_{1,2}^2 - \omega_{c+}^2 \cos^2\theta) - \omega_{1,2}^2 (\omega_{1,2}^2 - \omega_{c+}^2)\}}{\omega_{p+}^2 (\omega_{1,2}^2 - \omega_{c+}^2) \cos\theta} \right].
\end{aligned} \tag{161}$$

5.10. MULTI-COMPONENT PLASMA

So far we have assumed an ionospheric plasma which consists of two components, electrons and a single kind of ions. This is normally satisfied for the HF and VLF ranges throughout the outer ionosphere and for the ELF range above a height of

1000 to 1500 km. Below this height it is necessary to take into account the effects of ion species of different charge-to-mass ratios, i.e. H^+, He^+, O^+, etc.

In connection with the so-called electron and ion whistlers, we are now interested in the slow and fast Alfvén waves $(\theta = 0)$ in a multi-component plasma. The cold refractive indices of the slow and fast Alfvén waves may be expressed, referring to the first equation of (69), (11) and (81), as

$$n_{o,e}^{(0)2} = 1 - \frac{\omega_{p-}^2}{\omega(\omega \pm \omega_{c-})} - \sum_{+m} \frac{\omega_{p+m}^2}{\omega(\omega \mp \omega_{c+m})}$$

$$\simeq \mp \frac{\omega_{p-}^2}{\omega\omega_{c-}} - \sum_{+m} \frac{\omega_{p+m}^2}{\omega(\omega \mp \omega_{c+m})} \quad (\omega_{p-}^2 \gg \omega\omega_{c-}), \quad (162)$$

where ω_{c+1}, ω_{c+2}, ..., are the ion cyclotron frequencies of different species. One sees that the slow Alfvén wave has discrete frequency bands of propagation with resonances at the ion cyclotron frequencies ω_{c+1}, ω_{c+2}, ...

6. Specific Features of VLF and ELF Waves in the Outer Ionosphere

In the previous sections, formulas and features of the magnetoionic and magneto-hydrodynamic waves including plasma waves in the outer ionosphere have been presented over the whole frequency range.

In the present section we proceed to specific features of those waves in the VLF and ELF ranges together with some numerical examples for typical ionospheric conditions.

6.1. DISPERSION CURVES OF THE PURE AND MODIFIED ALFVÉN WAVES IN THE MAGNETO-
     SPHERE

Figure 4 shows the dispersion curves for 'whistler' and 'micropulsation' modes, i.e. for the pure and modified Alfvén waves, at a height around $(2-2.5)$ R_0 (R_0 is earth's radius). The refractive index now takes a complex value and the imaginary part represents Landau and cyclotron damping. The real part also has a small thermal correction. The ducted whistler mode or the fast Alfvén wave for parallel propagation accompanies electron cyclotron damping and has a resonance at the electron cyclotron frequency 0.26 Mc/sec (Equation (44)), while the ducted micropulsation or the slow Alfvén wave for parallel propagation accompanies ion cyclotron damping and has a resonance at the ion cyclotron frequency 144 c/sec (Equation (97)). Cyclotron damping is large near the cyclotron resonances. As the direction of propagation becomes oblique with respect to the geomagnetic field, Landau damping gradually increases, total damping becomes large and takes on a maximum at a certain angle somewhat below 90° (Equations (55), (107)). After that, the damping decreases rapidly and at an angle of exactly 90° there is no Landau and cyclotron damping. For the case of the whistler mode, this is just the lower hybrid resonance at a frequency of

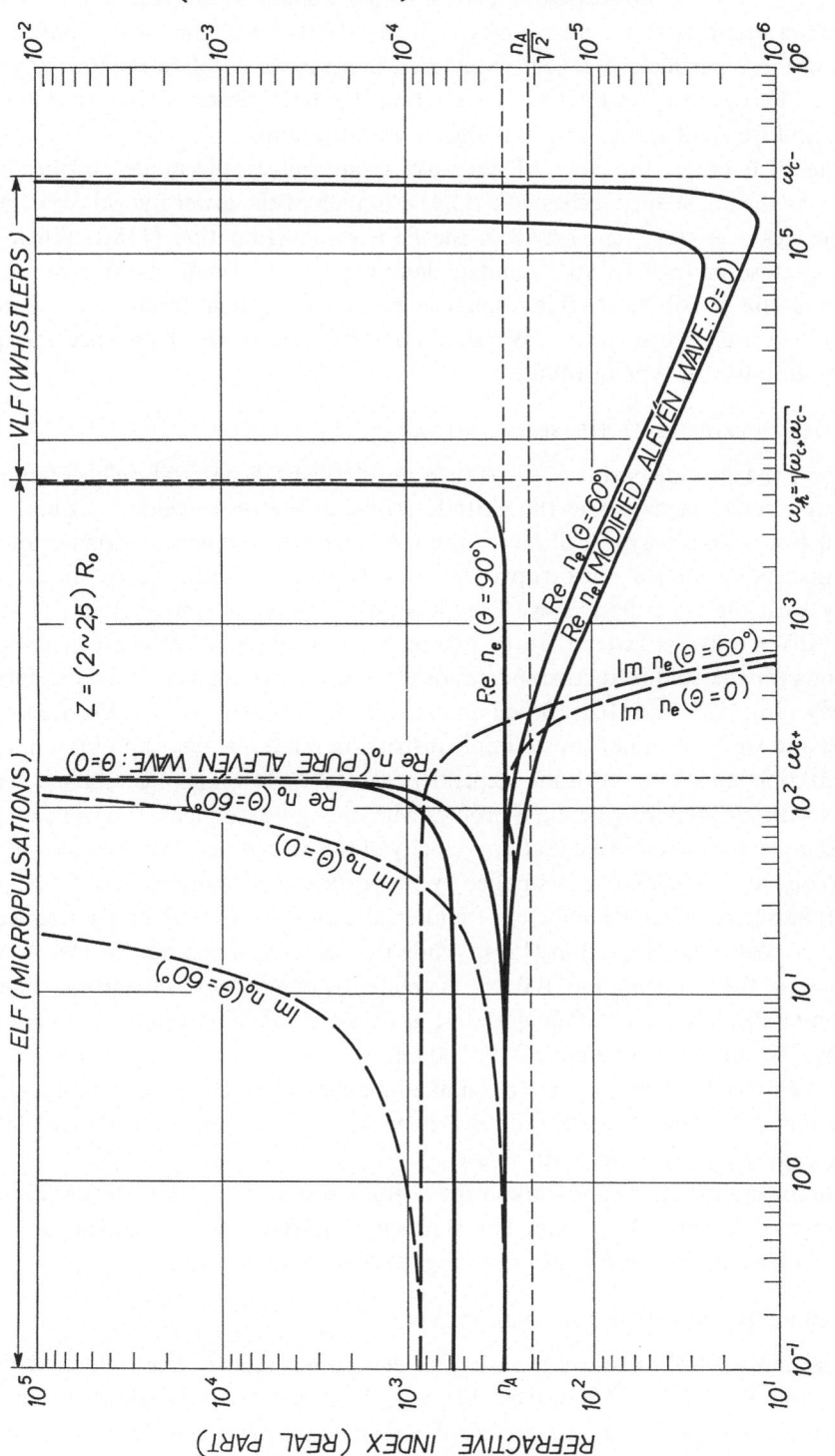

Fig. 4. Complex refractive indices for the pure and modified Alfvén waves.

$\omega \simeq \sqrt{(\omega_{c+} \omega_{c-})} = 6$ kc/sec (Equation (118)). In the vicinity of this resonance, the fast Alfvén wave becomes slow and couples with the electron plasma wave. Continuous VLF radiation is particularly favored and has a sharp lower cutoff at the lower hybrid frequency. This type of resonance or radiation has been observed by the Alouette satellite, and we shall see it soon in Alouette spectrograms.

For the ELF range, the slow Alfvén wave is an anisotropic mode and becomes very slow as the angle approaches 90°. It has a branch of the lower hybrid resonances at frequencies $\omega \simeq \sqrt{(\omega_{c+} \omega_{c-})} \cos \theta$ in the ELF range (Equation (118)). When the angle θ is close enough to 90°, Landau damping is very small. Continuous ELF radiation is thus possible near this resonance practically only at frequencies of order of about one cycle/sec or lower. We shall soon see one of the ELF spectrograms corresponding to this type of radiation.

6.2. ALTITUDE PROFILES OF THE SLOW AND FAST ALFVÉN WAVES

In Section 2 the altitude profile of the Alfvén speed v_A was discussed briefly (Figure 3). At the same time this expresses the altitude profile of the phase velocity of both the slow and fast Alfvén waves well below the ion cyclotron frequency. Because of the plasma dispersion and/or anisotropy, the velocity of the Alfvén waves varies considerably with the wave frequency. The slow Alfvén wave is slower than v_A, while the fast Alfvén wave is faster than v_A except in the vicinity of resonances. Explicit expressions for the cold refractive indices of these waves were given by rather simple forms (95), (96), (125) and (162). Based on the altitude profile of the Alfvén speed v_A in Figure 3, and with the aid of those simple dispersion relations, one can obtain plenty of information about the altitude dependence of the slow and fast Alfvén wave velocities without preparing detailed profiles with changing frequencies or angles.

For example, it is readily seen that the velocity of the ducted slow Alfvén wave ($\theta = 0$) of the frequency $\omega = 0.99 \, \omega_{c+}$ is smaller by one order of magnitude than v_A (Equation (96)) and is equal to the velocity of unducted slow Alfvén wave of the frequency $\omega \ll \omega_{c+}$ for $\cos \theta = 0.1$ (Equation (125)), while the velocity of the ducted fast Alfvén wave ($\theta = 0$) of the frequency $\omega = 100 \, \omega_{c+}$ is greater by one order of magnitude than v_A (Equation (162)). Then the Alfvén speed v_A is 1.5×10^6 m/sec at a height of 1000 km and $(3-4) \times 10^5$ m/sec at a height of 100000 km.

The Alfvén speed possesses a maximum of about 4×10^6 m/sec at a height of 3000 to 4000 km and at this height the ducted fast Alfvén wave with a frequency of 1–2 Mc/sec takes on a maximum of $(8-9) \times 10^7$ m/sec.

It is interesting to see the anomaly of the altitude profile of the Alfvén speed in the height range of 10000 to 30000 km. This is closely connected with the so-called 'knee' formation observed by whistlers in this height range.

6.3. WAVE EXCITATION IN THE MAGNETOSPHERE

In the magnetosphere, there are a variety of electromagnetic sources which radiate or excite the VLF and ELF waves. While a lightning flash provides a regular source for electron whistlers, the stream of energetic particles associated with mag-

TABLE II

Conditions for wave excitation

Type of radiation	s	v_0 Particle or beam velocity	$\omega = s\Omega + k \cdot v_{0\parallel}\cos\theta$ Emitted frequency	Type of beams and waves	Direction of movement of beams and waves	Type of instability
Čerenkov radiation	$s = 0$	$v = v_{0\parallel}\cos\theta$	$\omega = \mathbf{k}\cdot\mathbf{v}_0$	Beams + plasma mode	Co-streaming $v_{0\parallel} \gtrsim v$	Longitudinal
Normal Doppler effect	$s = 1, 2, 3, \ldots$	$v > v_{0\parallel}{}^a \cos\theta^b$	$\omega = \dfrac{\lvert s \rvert\,\Omega}{1 - (v_{0\parallel}/v)\cos\theta}$	Electrons + R-mode Ions + L-mode	Counter-streaming	Non-convective cyclotron
Cyclotron radiation Anomalous Doppler effect	$s = -1, -2, -3, \ldots$	$v < v_{0\parallel}\cos\theta$	$\omega = \dfrac{\lvert s \rvert\,\Omega}{(v_{0\parallel}/v)\cos\theta - 1}$	Electrons + L-mode Ions + R-mode	Co-streaming $v_{0\parallel} \gtrsim v$	Convective cyclotron

[a] $v_{0\parallel}$: projection of the velocity $\mathbf{v}_0$ to $\mathbf{B}_0$.
[b] θ: angle between $\mathbf{k}$ and $\mathbf{B}_0$

netic disturbances or trapped in the radiation belt emits Čerenkov or cyclotron radiation for the VLF and ELF ranges and amplifies the Alfvén waves through the process of Čerenkov or cyclotron instability as in the case of P_i1 or P_c1 pulsations. Interaction of the solar wind with the magnetosphere and turbulence on the periphery of the magnetosphere will generate toroidal and poloidal oscillations and consequently P_c5 and P_c2–4 micropulsations, respectively. Table II summarizes excitation conditions for these VLF and ELF waves.

In an equilibrium plasma, i.e. when the distribution function is Maxwellian, particles whose velocities are less than the wave phase velocity prevail, and absorb energy from the waves. Thus, the waves are damped. If there are more fast particles than slow ones in the neighbourhood of the wave phase velocity, the particles will give up energy to the wave rather than absorb it. This leads to an excitation or an instability for the waves. General excitation conditions are given by the relations in Table II, where the case $s = 0$ leads to Čerenkov radiation and the case $s \neq 0$ to cyclotron radiation.

A plasma beam travelling at an appropriate velocity in the direction opposite to the wave propagation encounters the Doppler-shifted wave whose rotating field direction matches the cyclotron motion of the particle: R-mode for electrons and L-mode for ions. This encounter results in the amplification of waves, the so-called 'non-convective cyclotron instability'. When a plasma beam infinitesimally exceeds the wave phase velocity in the same direction, the coupling is attained by electrons for L-mode and by ions for R-mode, causing the convective cyclotron instability. When the longitudinal plasma wave rather than the transverse electromagnetic wave exists, the longitudinal plasma instability occurs rather than the convective cyclotron instability. These, however, should be regarded only as the necessary conditions for the growth of waves. In case of collective interaction, for instance, it is required that the particle distribution function should satisfy some additional sufficient conditions.

6.4. GROUP DELAY TIME

In the present subsection, we attempt to introduce some kinetic considerations for study of sonagrams. The frequency dependence of the group delay time of the wave packets is determined by

$$\tau(\omega) = \int_s \frac{ds}{v_g(\omega, s)} = \frac{1}{c} \int_s n_g(\omega, s) \, ds, \qquad (163)$$

where v_g is the group velocity, n_g is the so-called group refractive index and ds is the path element. The group refractive index is related to the refractive index n by

$$n_g = n + \omega(dn/d\omega). \qquad (164)$$

For the case of satellite observations, the integral must be performed on a short path ending at the position of the satellite. It is normally postulated that the trajectory coincides with the geomagnetic field lines, since the group velocity vector $\mathbf{v}_g$, i.e. the energy flow, really is pressed towards the $\mathbf{B}_0$ vector. In addition, one may assume that ducts of elongated shape extending along the geomagnetic field lines are responsible for focussing signals.

For the ducted whistler mode or the fast Alfvén wave, the refractive index is given by Equation (44), while for the ducted micropulsation mode or the slow Alfvén wave, the refractive index is defined in Equation (97).

In the immediate vicinity of the ion or electron cyclotron frequency, the refractive index for the micropulsation or whistler mode takes on the form, instead of Equation (44) or (97),

$$\left.\begin{array}{c} n_o^2 \\ n_e^2 \end{array}\right\} = \sqrt[3]{\frac{\pi}{16\epsilon_\pm}}(1+i\sqrt{3})\sqrt[3]{\frac{\omega_{p\pm}^4}{\omega^4}\left(\frac{e^{-z^2\pm 1}}{\cos\theta}\right)^2}. \tag{165}$$

For unducted whistler and micropulsation modes, these formulas cannot be employed, and trajectories deviate from the magnetic field direction. However, once unducted modes whose wave vector makes an intermediate angle with the magnetic field approach resonances they are rapidly damped owing to large Landau damping. On the other hand, unducted modes whose wave vector is close enough to 90° with respect to the magnetic field are little damped near resonances.

For unducted whistler modes in the immediate vicinity of the lower hybrid resonance $\omega = \omega_{c-}\cos\theta$ $(0 < \theta \lesssim \pi/2)$, the refractive index can be written as

$$n_{ep}^2 = n_{ep}^{(0)2}\left(1 + \frac{i}{2}\sqrt{\frac{\pi}{2}}\cdot\frac{\epsilon_-^{-3/2}}{n_{ep}^{(0)}B_0}\cdot\frac{\omega_{p-}^2}{\omega^2}\cdot\frac{e^{-z^2_{0-}}}{\cos\theta}\right),$$

$$n_{ep}^{(0)2} = \sqrt{-\frac{B_0}{\epsilon_+ A_{1+} + \epsilon_- A_{1-}}}, \tag{166}$$

where the coefficients B_0, $A_{1\pm}$ are defined in Equations (15) and (84). The imaginary part represents electron Landau damping.

Substituting these expressions for the refractive indices, Equations (44), (97), (165), (166), into Equation (163) via Equation (164), we can obtain the complex group delay time. The imaginary part represents the integral damping or absorption of signals and this makes it possible, for instance, to evaluate cyclotron absorption near the upper cutoff frequency of nose whistlers and consequently to determine the velocity distribution function of electrons at the top of the wave trajectory (Liemohn and Scarf, 1962, 1964; Al'pert, 1967). Furthermore, the thermal correction to the group delay time itself (to its real part) is essential to determine the effective temperature in addition to the electron density at the wave trajectory apex (Guthart, 1965; Al'pert, 1967).

However, these formulas for the refractive indices are based on the assumption that the Maxwell distribution law is valid. Owing to the existence of high-energy particles in the outer radiation belt, a non-Maxwellian distribution $(\partial f/\partial v > 0)$ can be produced by large-scale electric fields. Then, the cyclotron damping term in Equation (44) or (97) may change sign and produce a cyclotron instability. The wave absorbs energy from particles and grows while propagating along the magnetic field lines and reflecting periodically at the magnetically conjugate points. This is the so-called 'convective cyclotron instability' which has been proposed for interpretation of a pearl-type micropulsation. We can estimate the amplification coefficient during the two-hop propagation of the signal along the field line by choosing an appropriate distribution function for energetic particles.

7. Some Top-Side Sounding Data

As is well known, ground-based observations have the serious limitation that they cannot provide information about the ionosphere at levels where the electron cloud reaches its maximum density corresponding to a critical probing frequency. Radio waves of frequency greater than this critical value penetrate the ionosphere and are not reflected to the earth.

The successful experiments carried out with the top-side sounder Alouette have made it possible to determine the electron density profiles at altitudes well above the level of maximum density. Furthermore, Alouette has intensified the interest in new observations of plasma resonances excited in the immediate vicinity of the satellite over the wide HF and VLF ranges. Other most interesting results obtained with the satellite Injun are new observations of the so-called electron and ion whistlers or ion cyclotron waves in the ELF range.

In this section, we consider some of these experimental data and give brief interpretations indicating how the theory discussed so far applies to the analysis of those results. Finally, a typical result of ground observations of micropulsations and/or ELF radiation will be included, on the one hand as evidence for the theory described above, and on the other as an indication to satellite-borne experiments in the future.

7.1. Magnetoionic Waves and Plasma Resonances in the HF Range

Figure 5 shows one of the top-side ionograms recorded by the satellite Alouette-II

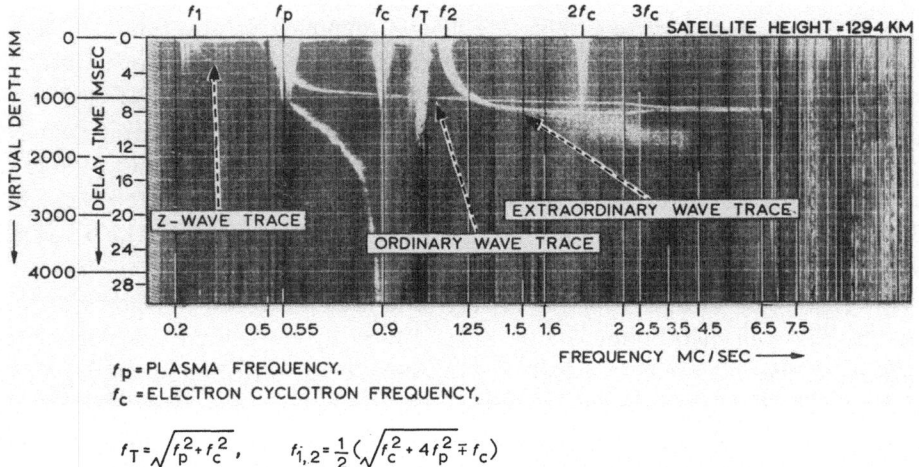

$$f_p = \text{PLASMA FREQUENCY},$$
$$f_c = \text{ELECTRON CYCLOTRON FREQUENCY},$$
$$f_T = \sqrt{f_p^2 + f_c^2}, \qquad f_{1,2} = \tfrac{1}{2}\left(\sqrt{f_c^2 + 4f_p^2} \mp f_c\right)$$

Fig. 5. Alouette ionogram with analysis, showing the various resonance spikes, wave traces, and the anomalous dispersion near the plasma frequency. Recorded at DRB, Ottawa, December 1, 1965.

(Kikuchi, 1967). Not only is its quality extremely good compared to Alouette-I ionograms but it clearly exhibits new fine structures for both the electron plasma waves (plasma resonance spikes) and the fast and slow electromagnetic waves (ordinary, extraordinary and Z-wave traces). These resonance spikes have been interpreted as a manifestation of electrostatic or plasma oscillations, and a zero or low group velocity matching concept has been employed in earlier investigations (Fejer and Calvert, 1964; Dougherty and Monaghan, 1966).

These resonance spikes and wave trace exits are determined by local properties in the immediate vicinity of the moving satellite and the duration of spikes, i.e. width and length of observed spikes, can be estimated in terms of the complex group refractive index whose imaginary part represents Landau and cyclotron damping. The group refractive index is derived from the refractive index which is described in a generalized form of the Appleton–Hartree equation discussed in Section 4. The use of the group refractive index and a low group velocity matching concept leads to the 'pseudo-Čerenkov effect' which arises from replacing the phase velocity or the phase refractive index with the group velocity or the group refractive index from the dispersion of a plasma medium. The concept seems convenient, since it implicitly contains the effects of moving sources, i.e. the Čerenkov concept is to some extent applied by using the group refractive index instead of the phase refractive index. In terms of a new language, the plasma resonance spikes are defined as a manifestation of the pseudo-Čerenkov effect of plasma waves or coupled plasma and slow electromagnetic waves.

Among the resonance spikes, the upper hybrid resonance seems most persistent in general (Equations (79), (80)), and other spikes are competitive with one another in duration mainly owing to the relative location of the plasma and cyclotron frequencies. However, the plasma resonance spikes seem to be the broadest, producing a wedge

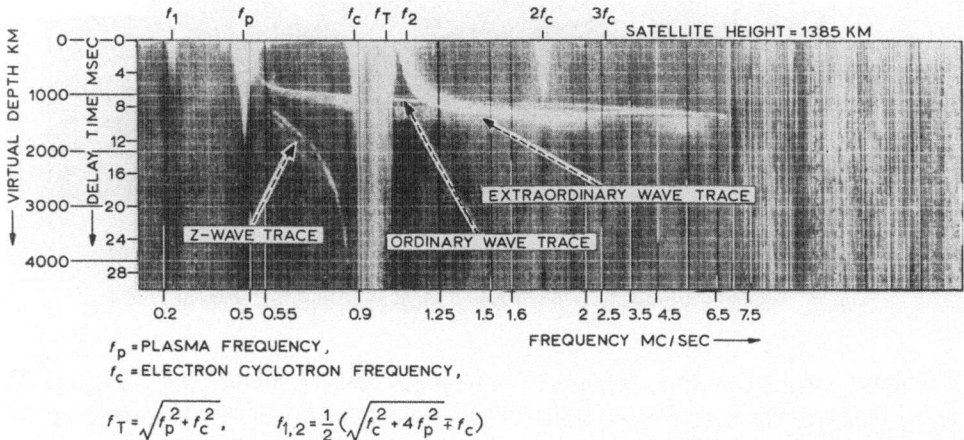

f_p = PLASMA FREQUENCY,
f_c = ELECTRON CYCLOTRON FREQUENCY,

$$f_T = \sqrt{f_p^2 + f_c^2}, \qquad f_{1,2} = \frac{1}{2}\left(\sqrt{f_c^2 + 4f_p^2} \mp f_c\right)$$

Fig. 6. Alouette ionogram with analysis, showing the various resonance spikes, wave traces, and the anomalous dispersion near the plasma frequency. Recorded at DRB, Ottawa, December 1, 1965, 1441 GMT.

shape as long as each resonance occurs far from other resonances, and they exhibit a more complex structure sometimes accompanying small side band spikes as shown in Figure 6 (Equations (23), (55)). This kind of spikes is also observed in the vicinity of the third wave trace exit f_1. This indicates the possibility of excitation of pure plasma waves near the wave trace exits by the enhancement of equivalent source dipoles due to the zero range echoes when the wave trace exits fall within the pseudo-Čerenkov zones for plasma waves.

Figure 6 also shows that the upper hybrid resonance is strongly pronounced, producing broad and long spikes by coupling with the cyclotron resonance when the two resonance frequencies almost overlap (Equation (61)).

While the ordinary and extraordinary wave traces which start from f_p and f_2, respectively, at the height of the satellite are quite familiar to us, the third-wave trace, the so-called Z-wave trace, exhibits a variety of behavior in the topside ionograms, though this kind of wave trace has occasionally been observed in the bottom side ionograms. The Z-wave is a slow wave, in contrast with the fast O- and X-waves, and this trace is asymptotic to the virtual Z-wave infinity spike at $f_+^2 = \frac{1}{2}\{f_p^2 + f_c^2 + [(f_p^2 + f_c^2) - 4f_p^2 f_c^2 \cos^2 \theta]^{1/2}\}$; this, however, does not come out in ionograms because of large Landau damping in the intermediate angles (Equation (166)). The Z-wave infinity spike should have been observed according to the Appleton–Hartree equation for a cold plasma.

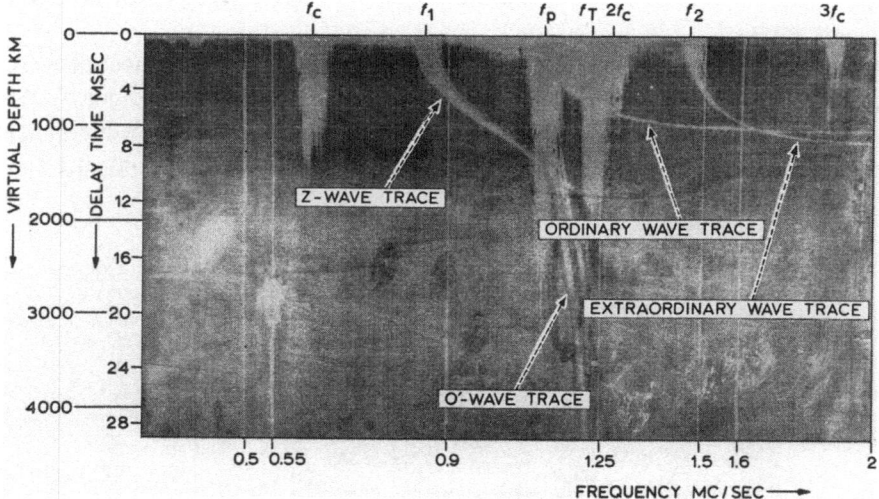

Fig. 7. Alouette II ionogram. Recorded at Singapore, July 23, 1966, 0932 GMT.

Figure 7 shows another Alouette-II ionogram. This is interesting owing to the appearance of the fourth-wave trace which starts from the plasma frequency at the height of the satellite, travels down, splitting from the ordinary wave, and approaches the Z-wave trace. An identification of this wave trace has been proposed as a ducted Z-wave transmitting and reflecting along the geomagnetic field lines (Calvert, 1966).

7.2. Lower hybrid resonance in the vlf range

Figure 8 shows a VLF spectrogram recorded by Alouette-I. We see the lower hybrid resonance noise band enhanced by atmospherics and/or whistlers. The noise band is from about 8 to 9 kc/sec in the lowest record and has a sharp lower frequency cutoff (Equations (79), (80), (117) to (123), (166)). The term 'lower hybrid frequency' is used to denote the resonance frequency for strictly perpendicular propagation. There is, therefore, no Landau damping exactly at the lower hybrid frequency. For the wave vector whose angle is less than 90°, however, the resonance frequency is higher than the lower hybrid frequency and there is Landau damping. Thus the resonance frequencies at which a resonance can occur have a natural lower frequency cutoff. For this reason, the noise band has a sharp lower frequency cutoff at the lower hybrid frequency. For angles close to 90° Landau damping is small. When the angle is decreased, Landau damping increases rapidly and resonances accompany absorption due to large Landau damping at intermediate angles. Thus the noise band is limited by Landau damping and has an upper frequency cutoff, though not so sharp as the lower frequency cutoff.

The occurrence of this noise band may be attributed to Čerenkov radiation due to the inflow of energetic particles into the outer ionosphere, since the Čerenkov condition is readily satisfied by possible energetic particles in the vicinity of the lower hybrid frequency. It is interesting to note that the plasma resonances observed in the HF range are a manifestation of the pseudo-Čerenkov effect, while the LHR emission is a true Čerenkov radiation by energetic particles.

Figure 9 also shows the lower hybrid resonance noise band, and a sharp lower

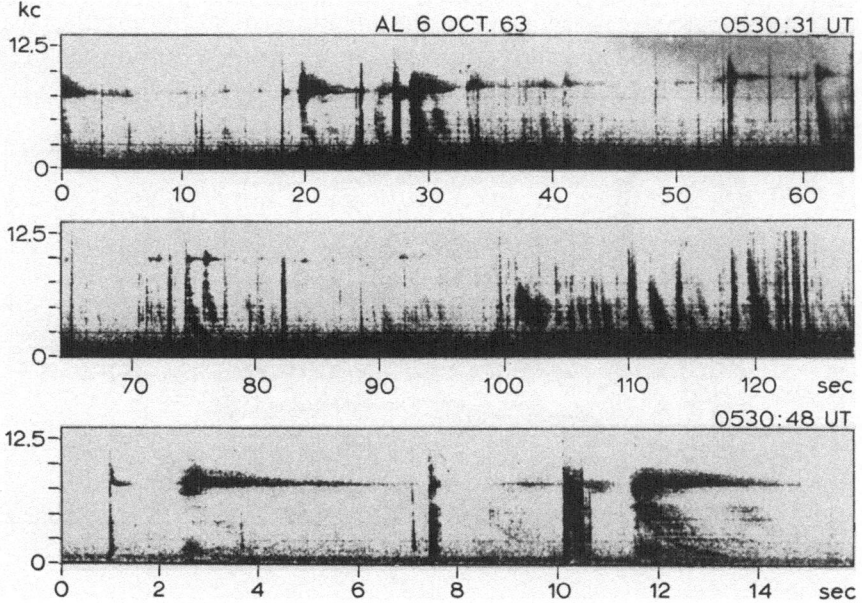

Fig. 8.   A VLF spectrogram of the lower hybrid resonance noise recorded by Alouette 1. Reproduced by courtesy of N. M. Brice and R. L. Smith.

Fig. 9.   A spectrogram of continuous VLF (LHR) radiation recorded by Alouette I.
Reproduced by courtesy of N. M. Brice and R. L. Smith.

frequency cutoff at the lower hybrid frequency is observed (Brice and Smith, 1965). The gradual decrease of the lower frequency cutoff indicates a spatial effect of the satellite location moving to higher latitudes. This indicates that the lower hybrid resonance is determined by local properties in the immediate vicinity of the satellite.

7.3. MAGNETOHYDRODYNAMIC WAVES IN THE ELF RANGE

Figure 10 shows an ELF spectrogram recorded by the satellite Injun (Gurnett *et al.*, 1965). The tail of the normal electron whistler in the ELF range is observed. It covers only a short distance from the source to the satellite (the electronic short-hop whistler's tail). Of great interest is the new appearance of the ion cyclotron wave or the pure Alfvén wave with left-hand circular polarization, i.e. the so-called ion whistler below the ion cyclotron frequency. The cutoff occurs due to ion cyclotron damping in the vicinity of the ion cyclotron frequency. These wave characters have been discussed in detail in Subsections 5.1, 5.2 and 5.10 (Equations (90) to (104), (162)).

In a multicomponent plasma, between each two adjacent ion cyclotron frequencies there is a frequency for which both modes of propagation are linearly polarized. These frequencies are the so-called crossover frequencies. A wave propagating in the ionosphere changes polarization at the altitude where the wave frequency is equal

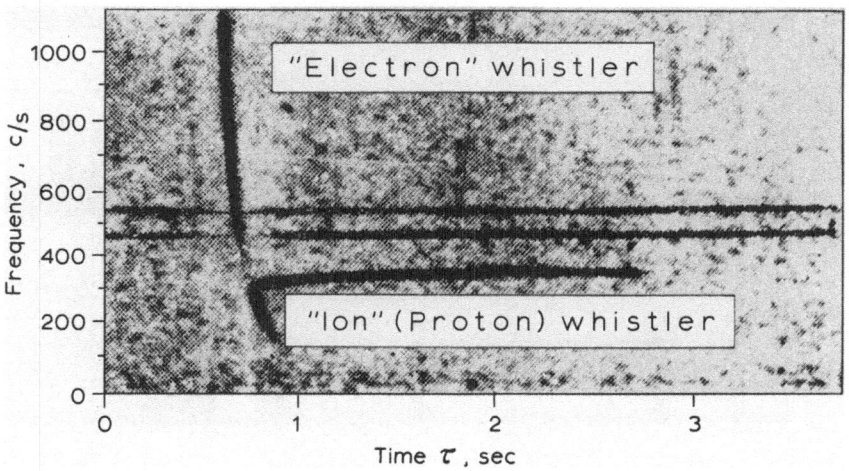

Fig. 10.   A spectrogram of ion and electron whistlers recorded by Injun.
Reproduced by courtesy of D. A. Gurnett.

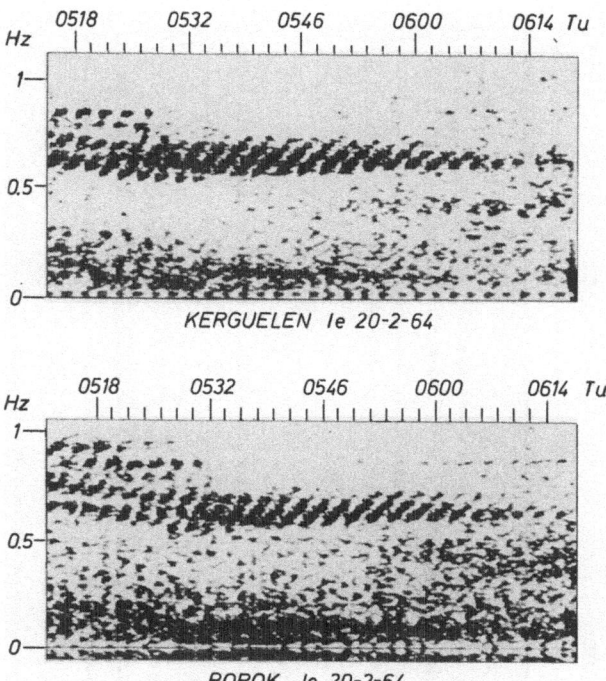

Fig. 11. Sonagrams of micropulsations recorded at magnetically conjugate points.
Reproduced by courtesy of R. Gendrin and V. A. Troitskaya.

to a crossover frequency. This polarization reversal provides the coupling mechanism between electron and ion whistlers. In Figure 10, in fact, weak cutoff of the electron whistler is observed while its coupling to the ion whistler takes place.

Figure 11 shows some sonagrams of discrete packets of micropulsations (Gendrin and Troitskaya, 1965). Many successive signals with a periodic structure are interpreted as wave packets propagating along the magnetic field lines and periodically reflecting between the magnetically conjugate points. As a rule, the frequency of the ion wave (slow Alfvén wave) increases with time (Equations (95) to (97), (107)) and the frequency variation rate is about 0.1 cycle per min. ELF waves of this type have been observed simultaneously in very extended regions in the northern and southern hemispheres. However, their amplitude decreases with latitude. This may be due to the fact that micropulsations generated in the magnetosphere reach the earth, being guided along the polar magnetic field lines, and then propagate along a hydromagnetic waveguide in the upper ionosphere, covering nearly the whole globe. While this is a typical ground observation of micropulsations, corresponding satellite observations are lacking.

Figure 12 shows continuous radiation of ELF waves at cycle frequencies and lower recorded at nearly magnetically conjugate points (Gendrin and Troitskaya, 1965). The radiation lasts for hours and is very similar in character at those points. This may be regarded as a branch of the lower hybrid resonances or ion plasma waves in the ELF range. Satellite observations of ELF radiation are also desired.

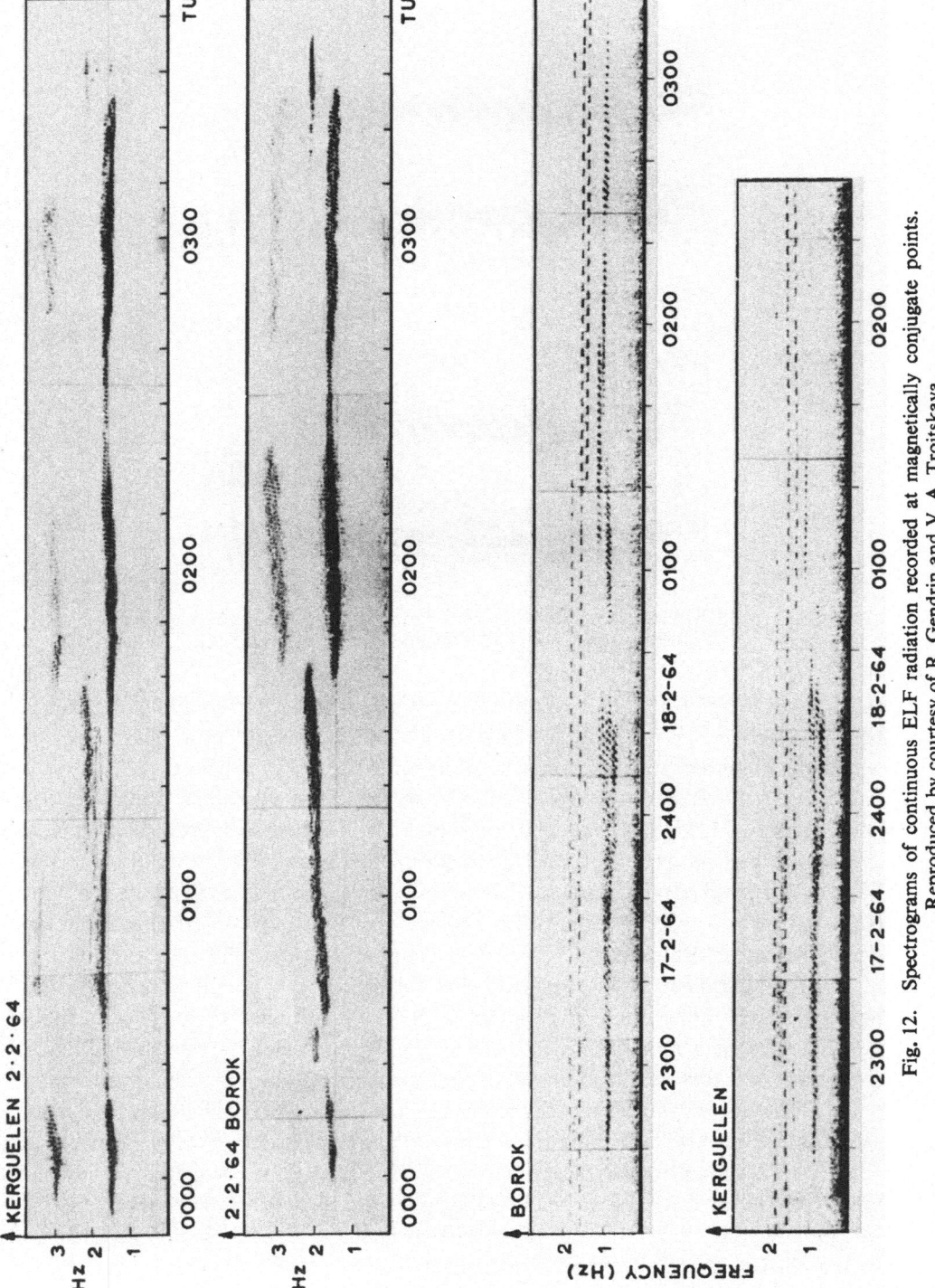

Fig. 12.   Spectrograms of continuous ELF radiation recorded at magnetically conjugate points.
Reproduced by courtesy of R. Gendrin and V. A. Troitskaya.

Summarizing the results of observations, we have two extreme cases of VLF and ELF spectrograms. One is the observation of a discrete type of wave packets and is concerned with ducted whistlers and micropulsations. This is rather a type of fast waves and is associated more or less with cyclotron damping or instability. The other is the case of continuous radiation for a long interval of time where the frequency band is normally limited by Landau damping. This is associated with Čerenkov radiation and waves are normally coupled with electron or ion plasma waves. The direction of radiation is almost perpendicular to the magnetic field, so that the waves are of unducted type.

We thus have seen, on a theoretical basis, a number of new satellite observations of magnetoionic, magnetohydrodynamic and plasma waves in the outer ionosphere over a whole range of frequencies. In conclusion it is pointed out that some satellite observations to be expected, for instance, micropulsations, ion plasma resonances and radiations in the ELF range, are still lacking and that these experiments should be conducted in the planning of satellite-borne measurements in the future.

Acknowledgments

The author wishes to express his thanks to Prof. W. Dieminger for his support in this study.

This report includes results obtained by the author at University of Toronto, Institute for Aerospace Studies. He wishes to express his thanks to Prof. G. N. Patterson for his support in this connection.

Alouette-II ionograms have been provided through the courtesy of the Defence Research Telecommunications Establishment of Canada. The author is particularly grateful to Dr. I. Paghis and Dr. E. S. Warren for making Alouette data available.

He also wishes to extend his thanks to Drs. W. Becker, G. Pfotzer, M. Siebert, and E. S. Warren for their interests in this study.

References

Al'pert, Ya. L.: 1967, *Space Sci. Rev.* **6**, 419.
Al'pert, Ya. L.: 1967, *Space Sci. Rev.* **6**, 781.
Brice, N. M. and Smith, R. L.: 1965, *J. Geophys. Res.* **70**, 71.
Barrington, R. E., Belrose, J. S., and Nelms, G. L.: 1965, *J. Geophys. Res.* **70**, 1647.
Calvert, W.: 1966, *J. Geophys. Res.* **71**, 5579.
Dougherty, J. P. and Monaghan, J. J.: 1966, *Proc. Roy. Soc. London* **289**, 214.
Fejer, J. A. and Calvert, W. J.: 1964, *J. Geophys. Res.* **69**, 5049.
Gendrin, R. and Troitskaya, V. A.: 1965, *Radio Sci.* **69D**, 1107.
Gurnett, D. A., Shawhan, S. D., Brice, N. M., and Smith, R. L.: 1965, *J. Geophys. Res.* **70**, 1665.
Guthart, H.: 1965, *Radio Sci.* **69D**, 1417.
Kikuchi, H.: 1963, *Proc. of the VIth International Symposium on Ionization Phenomena in Gases* (S.E.R.M.A., Paris) **3**, 13.
Kikuchi, H.: 1964, *Proc. of the Symposium on Quasi-Optics* **14**, Polytechnic Press, New York, p. 235.
Kikuchi, H.: 1964, University of Oxford, Dept. of Engineering Science, Oxford, Tech. Rept. No. 15, Contract No. AF 61 (514)-1183.
Kikuchi, H.: 1967, Presented at the Joint U.S.-Canadian URSI Meeting, Ottawa.

Kikuchi, H.: 1967, Presented at the Joint APS-SMF-CAP Meeting, Toronto.
Liemohn, H. B. and Scarf, F. L.: 1962, *J. Geophys. Res.* **67**, 1785.
Liemohn, H. B. and Scarf, F. L.: 1964, *J. Geophys. Res.* **69**, 883.
Lockwood, G. E. K.: 1963, *Canad. J. Phys.* **41**, 190.
Smith, R. L., Brice, N. M., Katsnfrakis, J., Gurnett, D. A., Shawhan, S. D., Belrose, J. S., and
    Barrington, R. E.: 1964, *Nature* **204**, 274.
Stix, T. H.: 1962, *The Theory of Plasma Waves*, McGraw-Hill, New York.
Storey, L. R. O.: 1953, *Phil. Trans. Roy. Soc (London)* **A246**, 113.
Storey, L. R. O. and Cerisier, J. C.: 1968, Groupe de Recherches Ionosphériques, Troitskaya, V. A.
    and Gul'elmi, A. V.: 1967, *Space Sci. Rev.* **7**, 689.
Warren, E. S.: 1962, *Canad. J. Phys.* **40**, 1692.
Warren, E. S.: 1963, *Nature* **197**, 636.
Warren, E. S.: 1963, *Canad. J. Phys.* **41**, 188.

EFFECTS OF ION-ATOM COLLISIONS ON THE
PROPAGATION AND DAMPING OF ION-ACOUSTIC WAVES*

H. K. ANDERSEN, N. D'ANGELO, V. O. JENSEN,
P. MICHELSEN and P. NIELSEN

Research Establishment Risö, Roskilde, Denmark

Abstract. Experiments are described on ion-acoustic wave propagation and damping in alkali plasmas of various degrees of ionization. An increase of the ratio T_e/T_i from 1 to approximately 3–4, caused by ion–atom collisions, results in a decrease of the (Landau) damping of the waves. At high gas pressure and/or low wave frequency a 'fluid' picture adequately describes the experimental results.

1. Introduction

A previous experiment (Andersen *et al.*, 1967) on shock formation in a Q-device indicated that, for equal ion and electron temperatures $T_i \simeq T_e$, Landau damping prevents shock formation by overcoming the sharpening effects of the nonlinearities. When the ratio T_e/T_i was made, however, as large as 3 or 4 by cooling the ions through ion–atom collisions, shock formation was observed.

In order to confirm and extend these results, we have next studied the effect of ion–neutral collisions on the propagation and damping of sinusoidal waves (rather than plasma 'pulses'). This new study is the subject of the present paper.

Propagation of ion-acoustic waves in a fully ionized plasma with equal ion and electron temperatures has been investigated by Wong *et al.* (1964) in the Princeton Q-3 device. In particular, it was found that the waves exhibit a strong (Landau) damping, with a ratio δ/λ between the attenuation length and the wavelength equal to ~ 0.4, independent of wavelength.

When neutral gas is added to the plasma, provided the wave frequency is sufficiently large and/or the neutral gas pressure not too high, the main effect of the neutral atoms should consist of cooling the ions, thereby varying the ratio T_e/T_i from 1 to 3 or 4. Ion cooling, in turn, is expected to reduce appreciably Landau damping.

At much higher neutral pressures and/or lower wave frequencies, on the other hand, one expects wave propagation and damping to be describable on the basis of a simple 'fluid' picture including the effect of ion–neutral collisions through some suitable collision term.

In intermediate ranges of neutral gas pressure and/or wave frequency, two additional effects may be expected to appear. One is a transition from Landau damping to 'viscous' damping, as experimentally demonstrated by Motley and Wong (1964); the other, the necessity of including, at still moderate frequencies, an ion–atom collision term in the Vlasov equation.

We find it convenient to present our experimental results according to the pressure range of the neutral gas, namely low pressure, high pressure, and intermediate pressure.

* This is a reprint of a paper published in *Phys. Fluids* **11** (1968) 1177.

It will, of course, be realized that a more relevant parameter is the ratio ω/v_{in} between the wave frequency and the ion–neutral collision frequency. For instance, an expression for v_{in} is given in Joos (1950).*

The paper is organized as follows. Section 2 describes the experimental setup, Section 3 presents the experimental results, Section 4 contains the discussion, and Section 5 the conclusions.

2. The Experimental Setup

The experiments were performed in the Q-device at the Research Establishment Risö. This Q-device is similar in construction to other alkali-plasma sources described in the literature. Its novel features have already been briefly described in Andersen *et al.* (1967).

The plasma is produced by surface ionization of a beam of cesium or potassium atoms on a hot (~ 2500 K) tantalum plate and confined radially by a magnetic field of intensity up to 10 000 G. The plasma column is ~ 125 cm long and 3 cm in diameter. It is terminated by a second tantalum plate which can also be heated to ~ 2500 K and bombarded with an alkali-metal beam similar to that of the first plate. Heating of the plates is accomplished as usual by electron bombardment. Separate pumping systems for the plasma region and the plate bombardment regions allow introduction of inert gases in the plasma. The neutral gas pressure in the column has been varied in the range $\sim 10^{-6}$ to $\sim 5 \times 10^{-2}$ mm Hg. The plasma density, in the range 10^{10} cm^{-3}–10^{12} cm^{-3}, is measured by means of Langmuir probes which are movable either across or along the column and the exposed tips of which are, typically, 2 mm long and 0.2 mm in diameter.

The ion-acoustic waves are excited by a plane grid inserted normally to the B lines 30 cm from one tantalum plate. The grid consists of parallel wires 2.5×10^{-3} cm in diameter, spaced 3×10^{-2} cm. The grid is biased at a small negative voltage (5–6 V) and supplied with a sinusoidal voltage of 2–3 V peak to peak. The ion-acoustic waves generated by the grid are detected by means of either a similar grid or a Langmuir probe movable along the plasma column. This technique is entirely similar to that described in Wong *et al.* (1964) and allows phase and amplitude measurements. It should be noted, however, that in the present experiments we have tried to avoid the long and tedious task of making the 'downstream' and 'upstream' measurements of Wong *et al.*, which were necessary there in order to obtain a precise measurement of the drift velocity of the plasma column from one end plate to the other. As each hot tantalum plate is now provided with its own oven, conditions can be arranged to reduce the plasma drift to small values.

3. Experimental Results

As already remarked in the Introduction, we find it convenient to divide our experi-

___

* On pp. 559 and 585 we find: $v_{in} = \sqrt{2}\sigma N_g(1.13)(2\chi T_i/m_i)^{1/2}$, where σ is the collision cross section, N_g the neutral-gas density, m_i the ion mass, and T_i the ion temperature.

mental results into three parts, referring to the low-pressure, high-pressure, and intermediate-pressure ranges.

A. LOW PRESSURES

The data in this subsection were all obtained by using argon as the neutral gas, with the pressure never exceeding $\sim 5 \times 10^{-4}$ mm Hg. Both cesium and potassium plasmas were employed. A typical 'run' consisted of fixing the wave frequency and performing measurements of the wavelength λ and of the attenuation length δ for several neutral gas pressures in the range $\sim 10^{-6}$ to $\sim 5 \times 10^{-4}$ mm Hg. Results obtained at several

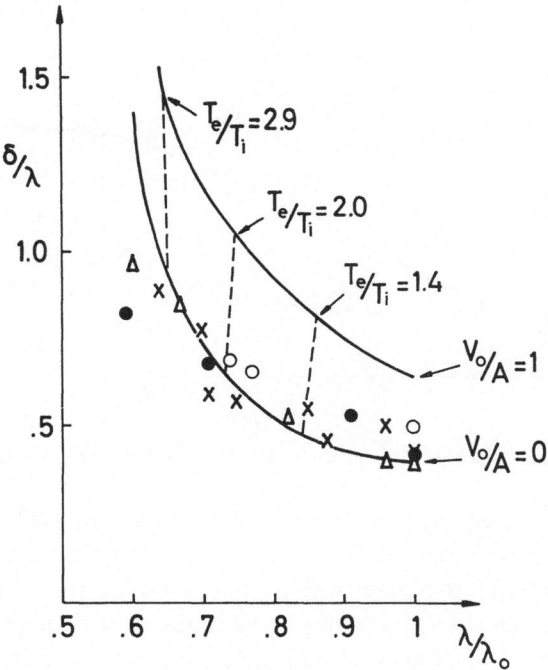

Fig. 1. The quantity δ/λ vs. λ/λ_0 at several frequencies for both cesium and potassium [($\times$) K, $f = 25$ kc/sec; $\bigcirc$ K, $f = 40$ kc/sec; $\bullet$ Cs, $f = 20$ kc/sec; $\triangle$ Cs, $f = 5$ kc/sec]. The two solid lines are the predictions of the collisionless theory for $V_0/A = 0$ and $V_0/A = 1$.

frequencies in cesium and potassium plasmas are shown in Figure 1. The ratio δ/λ is plotted as a function of λ/λ_0; λ_0 being the value of λ for pressures $\lesssim 10^{-5}$ mm Hg ($T_i = T_e$). The two solid lines marked $V_0/A = 0$ and $V_0/A = 1$, show the predictions of the collisionless theory for the two extreme cases in which (a) no drift of the plasma column is present and (b) this drift is equal to the ion-acoustic speed. V_0 is the drift velocity of the column and $A^2 = 2\kappa T_i/m_i$. Values of T_e/T_i at three points along the solid curves are also indicated in the figure.

It may be seen from Figure 1 that, in this pressure range, the main effect of ion–atom collisions is ion cooling. The corresponding variation of T_e/T_i from 1 to ~ 3 has the effect of reducing Landau damping, i.e., increasing δ/λ from 0.4 to ~ 1. It also seems clear that, when sufficient care is devoted to 'balancing' the plate temperatures and the oven fluxes at both ends of the plasma column, the condition is attained of a small drift from the exciter to the receiving grid.

B. HIGH PRESSURES

The data in this subsection were obtained by using a potassium plasma and argon as the neutral gas. The argon pressure was always kept in the range $\sim 10^{-3}$ to $\sim 5 \times 10^{-2}$ mm Hg. The data of Figure 2 refer to an argon pressure of 5×10^{-3} mm Hg, the wave frequency having been varied between 0.2 and 50 kc/sec. Over this frequency

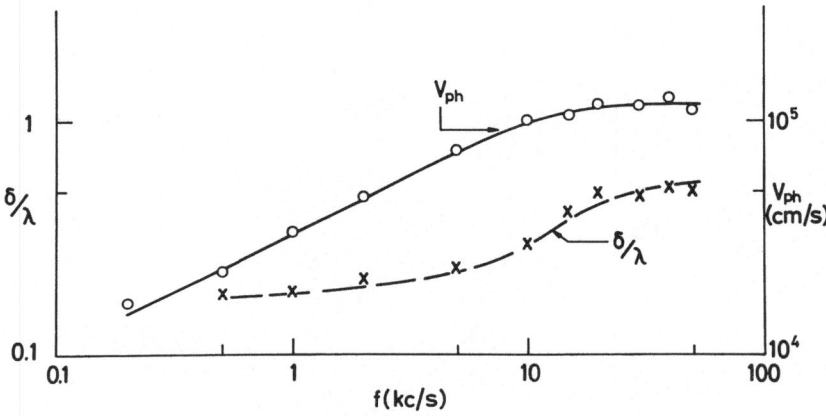

Fig. 2. v_{ph} and δ/λ vs. frequency potassium plasma, $p(A) = 5 \times 10^{-3}$ mm Hg.

range, two different regimes are covered. At frequencies $f \gtrsim 20$ kc/sec we approach the collisionless regime. At lower frequencies, where the plasma behavior should well be described by a 'fluid' treatment with a suitable ion-neutral collision term, we find that: (a) the phase velocity v_{ph} is proportional to the square root of the frequency and (b) the ratio δ/λ reaches, for very low f, a saturation value somewhat smaller than 0.2. It will be seen in the discussion that both features (a) and (b) are predicted by the 'fluid' treatment. This treatment predicts that $\delta v_{\mathrm{ph}} = 2C^2/v_{\mathrm{in}}$, where $C^2 = \kappa (T_i + T_e)/m_i$. The measured dependence of δv_{ph} on the neutral gas pressure (proportional to v_{in}) is shown in Figure 3, for $f = 1$ kc/sec. Figure 4 shows δv_{ph} vs. v_{ph} for neutral gas pressures of 1×10^{-3} and 5×10^{-3} mm Hg. It is evident that for small v_{ph} the product δv_{ph} tends to be nearly constant. The 'fluid' analysis shows that, for a fixed (very low) frequency, $v_{\mathrm{ph}} \sim v_{\mathrm{in}}^{-1/2}$ and $\delta/\lambda \simeq 1/2\pi$. Results obtained at a frequency $f = 1$ kc/sec in the pressure range 10^{-3} to $\sim 4 \times 10^{-2}$ mm Hg are shown in Figure 5.

C. INTERMEDIATE PRESSURES

At intermediate gas pressures ($p \sim 5 \times 10^{-4}$ mm Hg) and plasma densities $n \gtrsim 5 \times 10^{11}$ cm^{-3}, we expect to observe in a plot of v_{ph} and δ/λ vs. frequency the features which are typical of a transition from pure Landau damping to 'viscous' damping

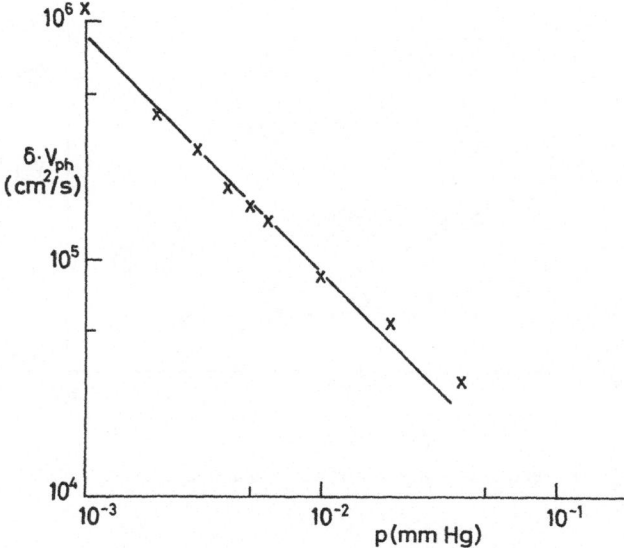

Fig. 3. δv_{ph} vs. argon pressure. Potassium plasma, $f = 1$ kc/sec.

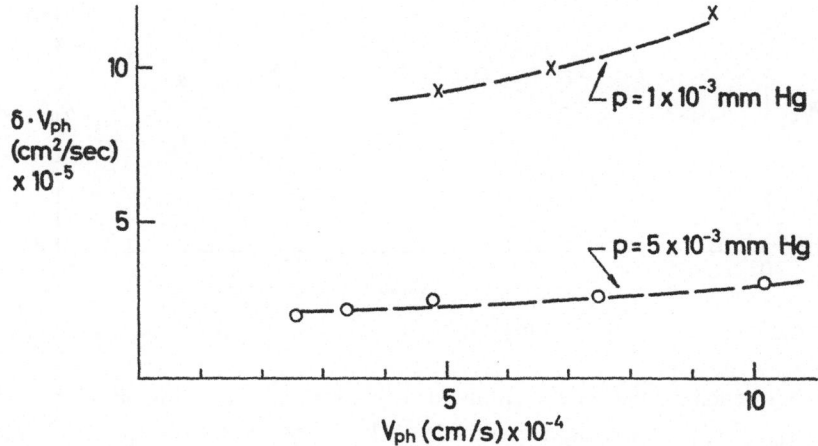

Fig. 4. δv_{ph} vs. v_{ph} for $p(A) = 1 \times 10^{-3}$ and 5×10^{-3} mm Hg. Potassium plasma.

(ion–ion collisions), in the manner of Motley and Wong (1964). That this may have been the case in our measurements can be inferred from the data shown in Figure 6 ($p = 5 \times 10^{-4}$ mm Hg and $n \sim 5 \times 10^{11}$ cm^{-3}). This set of data is more complicated than that in Figure 2. The broad maximum in δ/λ near $f \sim 5$ kc/sec is of the type expected on the basis of the Motley–Wong effect. If this interpretation is correct, the maximum should shift to a higher frequency when n is increased, while keeping the neutral gas pressure constant. A shift of the maximum to $f \sim 20$ kc/sec was in fact observed when increasing the density by approximately a factor of 4.

A second feature is also present in the data of Figure 6. This is a slight increase of δ/λ with frequency, for $f \gtrsim 20$–30 kc/sec, with a tendency for δ/λ to saturate at much

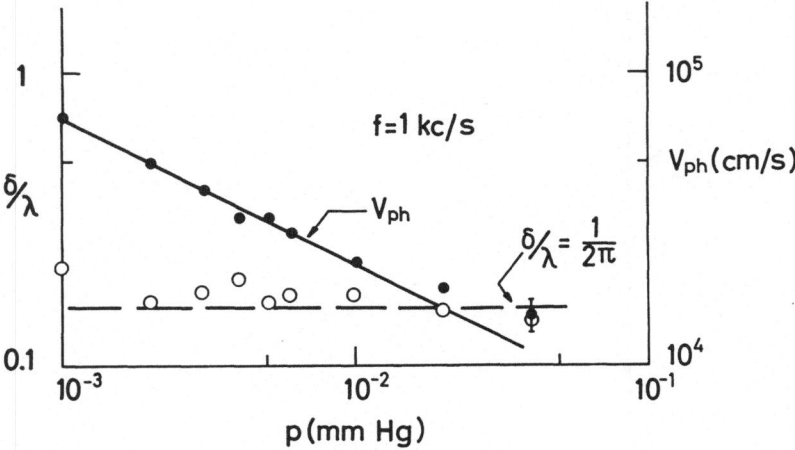

Fig. 5. v_{ph} and δ/λ vs. argon pressure. Potassium plasma, $f = 1$ kc/sec.

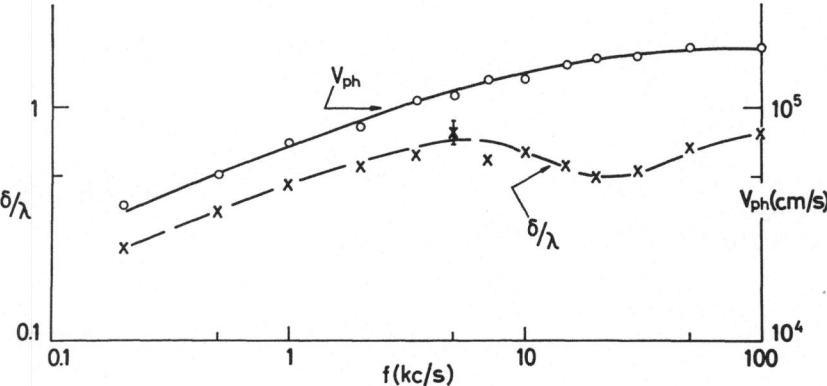

Fig. 6. v_{ph} and δ/λ vs. frequency. Potassium plasma ($n \sim 5 \times 10^{11}$ cm^{-3}), $p(A) \simeq 5 \times 10^{-4}$ mm Hg.

higher frequencies. One may account for this observation by introducing a collision term $-v_{in}(f-f_0)$ on the right-hand side of the Vlasov equation, to include the effect of ion–neutral collisions.

4. Theory and Discussion

In discussing our experimental results it will also be convenient to retain the broad separation of our data into the three main groups of Section 3. First we consider the high-frequency/low-pressure regime. Here we may use the result of the collisionless theory already employed in the discussion of the experiments of Wong *et al.* (1964). One writes

$$Z'\left[\frac{1}{A}\left(\frac{\omega}{K} \pm V_0\right)\right] = 2\frac{T_i}{T_e}, \tag{1}$$

where $Z'(x)$ is the derivative of the plasma dispersion function (Fried and Conte, 1961) and $A^2 = 2\kappa T_i/m_i$. We have solved Equation (1) numerically for several values of the ratio T_e/T_i, and for the two extreme cases of $V_0/A = 0$ and $V_0/A = 1$. The full lines in Figure 1 are the results of these calculations. It has already been remarked in Section 3 that, by a careful adjustment of the plate temperatures and the oven fluxes, we were apparently successful in reducing the drift of the plasma column to such an extent that our experimental points in Figure 1 would cluster around the line $V_0/A = 0$.

For the case in Section 3C, namely the 'intermediate' pressure regime, a good fit with experiment is obtained, in the high-frequency region, by including on the right-hand side of the Vlasov equation a collision term of the type $-v_{in}(f-f_0)$. Here f and f_0 stand for the actual and undisturbed ion distribution functions, respectively. In this case one obtains the same result as for the case of no collisions, except for a replacement of the wave angular frequency ω with $\omega + iv_{in}$. In particular, it follows that the ratio δ/λ is now given by

$$\delta/\lambda \approx \left(\frac{\delta}{\lambda}\right)_{\omega\to\infty} \frac{\omega/v_{in}}{\omega/v_{in} + 2\pi(\delta/\lambda)_{\omega\to\infty}}.\tag{2}$$

A 'fluid' picture with a suitable collision term will, on the other hand, be appropriate for the description of the high-pressure/low-frequency results. Here, theoretical results for comparison with the observations are most easily obtained by linearization of

$$\frac{\partial n}{\partial t} + \frac{\partial}{\partial x}(nv) = 0,$$

$$nm_i\frac{\partial v}{\partial t} + nm_iv\frac{\partial v}{\partial x} + \kappa(T_i + T_e)\frac{\partial n}{\partial x} = -v_{in}nm_iv,\tag{3}$$

which describe the ion motion in the presence of a uniform and stationary neutral gas background. With perturbations of the type $e^{i(Kx-\omega t)}$, with real ω and complex K, $K = K_r + iK_i$, and $C^2 = \kappa(T_i + T_e)/m_i$, one immediately obtains the following relations:

$$v_{ph}^2 = \frac{2C^2}{1 + [1 + (v_{in}/\omega)^2]^{1/2}}\tag{4}$$

and

$$\delta v_{ph} = 2C^2/v_{in}, \quad (\delta = 1/K_i).\tag{5}$$

From Equations (4) and (5), for $v_{in}/\omega \gg 1$, it follows

$$v_{ph} \simeq 2\pi^{1/2}C(f/v_{in})^{1/2}\tag{6}$$

and

$$\delta/\lambda \simeq 1/2\pi,\tag{7}$$

f and λ being the wave frequency and the wavelength. In Section 3, Equations (5), (6), and (7) have been used for a direct comparison with the experimental results.

5. Conclusions

These experiments on wave propagation and damping in a partially ionized alkali-plasma indicate that in the low-pressure/high-frequency regime a collisionless treatment is adequate, provided the effect of ion cooling by ion–neutral collisions is taken into account. A decrease of (Landau) damping is observed as the ratio T_e/T_i increases from its normal value of 1 to 3 or 4. Trust in the collisionless theory may extend so far as to consider the results in Figure 1 as an actual measurement of the ion temperatures. Note also that the present findings support our previous conclusions (Andersen *et al.*, 1967) concerning the role played by Landau damping in shock formation.

The results in the high-pressure/low-frequency regime seem entirely accountable for in terms of a very simple fluid treatment, with a collision term included to represent ion–atom collisions. Notice that a more sophisticated treatment (Kahn, 1965), including neutral-gas motion, in this case seems unnecessary.

In conclusion we may remark that the present experiment is also relevant as far as Q-machine technique is concerned. It shows that, through ion–atom collisions, the ratio T_e/T_i can be varied in a controllable fashion, a fact of interest to future wave experiments.

Acknowledgments

We wish to thank Mogens O. Nielsen and Borge Reher for their help at several phases of the work.

References

Andersen, H. K., D'Angelo, N., Michelsen, P., and Nielsen, P.: 1967, *Phys. Rev. Letters* **19**, 149.
Fried, B. D. and Conte, S. D.: 1961, *The Plasma Dispersion Function*, Academic Press, New York.
Joos, G.: 1950, *Theoretical Physics*, Hafner, New York.
Kahn, D.: 1965, *Phys. Fluids* **8**, 399.
Motley, R. W. and Wong, A. Y.: 1964, in *Proceedings of the Sixth International Conference on Ionization Phenomena in Gases* (ed. by P. Hubert and E. Crémien), Bureau des Editions, Centre d'Etudes Nucléaires de Saclay, Paris, Vol. III, p. 133.
Wong, A. Y., Motley, R. W., and D'Angelo, N.: 1964, *Phys. Rev.* **133**, A436.

EFFECTS OF FIELD-ALIGNED CURRENTS ON THE
STRUCTURE OF THE IONOSPHERE*

L. P. BLOCK and C.-G. FÄLTHAMMAR

Division of Plasma Physics, The Royal Institute of Technology, Stockholm, Sweden

Abstract. The influence of electric currents along the magnetic field lines on the high-latitude F-layer ionosphere is investigated theoretically. It is shown that a current of either sign leads to a reduction of the total electron content and the maximum density, if the charge carriers have to be produced in the ionosphere. Low-energy precipitation leads to a corresponding increase. Extremely low densities in the topside ionosphere may be reached for ion currents of the order of 10^{-6} amp/m² or electron currents of 10^{-3} amp/m². Observational evidence exists for currents at least of the order of 10^{-6} amp/m². It is suggested that the ionospheric trough, the F-layer storm, and other similar effects may be explained by field-aligned currents or plasma diffusion, which is also quantitatively accounted for by the present theory.

1. Introduction

With the growing knowledge of magnetospheric phenomena it has become increasingly clear that the ionospheric current systems, at least those associated with auroras and magnetic storms, do not flow entirely in the ionosphere but are also connected with the magnetosphere by field-aligned currents. The existence of such currents were first invoked by Birkeland (1913) and Alfvén (1939, 1940, 1950) and has more recently been discussed by Boström (1964, 1968) and by Cummings and Dessler (1967).

It is the purpose of the present paper to show that such field-aligned currents can have important effects on the structure of the ionosphere. It is shown that these effects can be quite pronounced at the current densities that can be expected. It is suggested that they may at least partly account for some of the hitherto unexplained F-layer anomalies. Only a qualitative discussion and analysis of idealized cases will be given here. However, the simplified model is chosen so as to contain the essential physical features of the real situation. Details and extensions will be given in later papers.

The horizontal parts of the ionospheric current systems flow mainly in the E region, where the Pedersen and Hall conductivities have their largest values (Boström, 1964). We shall consider the region above this level. There the total field-aligned current is essentially constant along a magnetic flux tube. However, the ratio at which ions and electrons contribute to this current need not be everywhere the same. At the lower end of each flux tube where the ratio is determined by the mobilities, the field-aligned current is carried mainly by electrons. At higher levels gravity and pressure gradients also play a role, and the ratio can be quite different. If the current ratio at the top end differs from that of low levels, certain important consequences occur.

We shall first discuss these consequences qualitatively in terms of four elementary cases illustrated in Figure 1. (Any real situation is some combination of two of these elementary cases.)

* This is a reprint of a paper published in *J. Geophys. Res.* **73** (1968) 4807.

If at the top end, the current were carried by *upward-moving ions* only, the field-aligned ion current would have a positive height-integrated divergence (Figure 1a) in the region considered, and the electron current a negative divergence, the total current having zero divergence. Hence the ion *flux* and the electron *flux* would both have positive and equal divergences. For a steady state to be possible this requires a corresponding deficiency in the height-integrated recombination rate relative to the

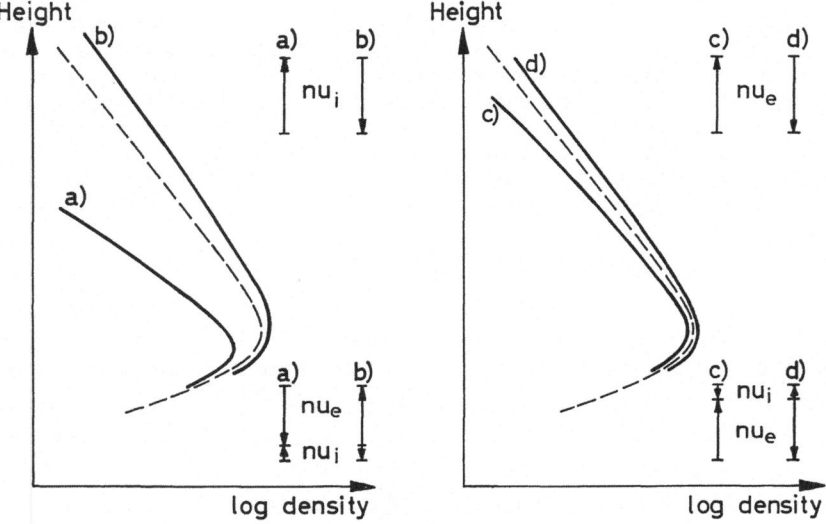

Fig. 1.   Density change due to field-aligned currents in four elementary cases. Full curves show actual density; dashed curves show density in absence of current. (a) Upward ion flow, no precipitation; (b) ion precipitation, no upward flow; (c) upward electron flow, no precipitation; (d) electron precipitation, no upward flow.

ionization rate. Now, the recombination rate is essentially proportional to the electron density, and the ionization rate is independent of it. Therefore the steady state has to be one where the *electron (and ion) density is reduced*.

If the topside current were instead carried by *upward-moving electrons only* (Figure 1c), the same kind of argument would again lead to the conclusion that the steady state would be one of reduced electron concentration. However, with a given total current the divergences and hence the electron concentration reduction is much less in case 1c than in case 1a (because the difference between topside and bottomside current components is smaller).

As illustrated in Figures 1b and d, down-flowing particles lead to a *density increase* instead, and again the effect is largest when ions carry the topside current.

In any real situation particles of both signs contribute, which means a combination of either 1a and 1d or 1b and 1c. Thus two opposing effects are involved, but the net effect does not vanish except when the current ratios are exactly constant along each flux tube, including both ends.

2. Mathematical Model

To demonstrate the essential physical effects we consider a simplified one-dimensional model. It is rather similar to the one used by Yonezawa (1959), the most important difference being that field-aligned currents are included in the present model. For simplicity it is assumed that transverse currents flow only above and below the region considered (the F region including the topside). As the horizontal currents serve only as external sources, their position is unessential for the model, and for mathematical convenience we can put them at $z = \pm\infty$ (the z axis is vertical, positive upward). As the auroral and polar ionosphere is of main interest we may consider the magnetic field to be vertical. Further we assume for simplicity that (a) the gravitational field is height-independent and the magnetic field is homogeneous; (b) the neutral atmosphere consists only of monoatomic oxygen with constant temperature T_n; (c) the ionosphere consists of electrons with constant temperature T_e and monoatomic oxygen ions with constant temperature T_i; (d) the rate of ionization is given by the Chapman distribution, (e) the rate of recombination is proportional to the product of the neutral density and the electron concentration; i.e., a recombination coefficient inversely proportional to the electron concentration is assumed.

In the present treatment we will only consider stationary states. Inertia terms are neglected. Then the basic equations that determine the electron concentration n are the following:

$$kT_i \frac{dn}{dz} = -n\left[m_i g - eE + v_{in} m_i u_i\right] + P_{ie} \tag{1}$$

$$kT_e \frac{dn}{dz} = -n\left[m_e g + eE + v_{en} m_e u_e\right] - P_{ie} \tag{2}$$

$$q = q_0 \exp\left[-\frac{z}{H_n} - b e^{-z/H_n}\right] \tag{3}$$

$$r = a_r n n_n \tag{4}$$

$$n_n = n_0 e^{-z/H_n} \tag{5}$$

$$i = ne(u_i - u_e) \tag{6}$$

$$\frac{d}{dz}(nu_i) = \frac{d}{dz}(nu_e) = q - r. \tag{7}$$

Equations (1) and (2) describe the momentum balance of ions and electrons, respectively. Subscript i refers to ions, e to electrons, and n to neutral particles. E is the electric field strength, and v are collision frequencies. P_{ie} represents the friction between ions and electrons.

Equation (3) gives the rate of ionization q, and Equation (4) gives the rate of recombination r.

Equation (5) is the barometric formula for the neutral atmosphere, H_n being the scale height. The current density i is given by (6). Particle conservation is taken care of by (7).

The choice of the level $z=0$ is arbitrary. We have defined it so that n_0 in Equation (5) has the value 10^{15} atoms/m^3. This means that $z=0$ at an altitude of approximately 300 km.

We also require an expression for the height dependence of the collision frequencies. We assume for simplicity that v_{in} and v_{en} are proportional to the neutral density and have the values v_{i0} and v_{e0} at $z=0$.

The Equations (1)–(7) may be combined to give

$$\frac{d^2 n}{d\zeta^2} - \frac{\kappa}{\zeta}\frac{dn}{d\zeta} + \left(\frac{\kappa}{\zeta^2} - 1\right) n = - Ne^{-\gamma\zeta},$$ (8)

where

$$\zeta = \beta e^{-z/H_n}$$ (9)

$$\beta = H_n \left[\frac{a_r n_0}{k(T_i + T_e)}(v_{i0}m_i + v_{e0}m_e)\right]^{1/2}$$ (10)

$$\gamma = b/\beta$$ (11)

$$\kappa = \frac{T_n}{(T_i + T_e)}$$ (12)

$$N = \frac{q_0}{a_r n_0}.$$ (13)

For a quantitative analysis we now choose the following values for the parameters involved: $n_0 = 10^{15}$ atoms m^{-3}; $q_0 = 10^8$ m^{-3} sec^{-1}; $T_n = 1500$ K; $T_i = 2000$ K; $T_e = 2500$ K; $v_{i0} = 0.8$ sec^{-1}; $v_{e0} = 40$ sec^{-1}; $\beta = 0.5$; $b = 0.1$; $\gamma = 0.2$; $\kappa = 1/3$; $N = 10^{12}$ m^{-3}.

Other values of β and γ have also been used, but the results are essentially similar to those discussed below.

The current density i does not enter in the differential Equation (8) for the density n. However the boundary conditions can be expressed in terms of the electron and ion currents at $z=\infty$ in a way that will be discussed below.

3. Properties of the Solution

The solution of (8) satisfying the boundary condition $n=0$ at $=\pm\infty$ can be written in the following form

$$n = C_0 \zeta^{2/3} I_{-1/3}(\zeta) + n_p(\zeta) + C\zeta^{2/3} K_{1/3}(\zeta),$$ (14)

where $I_{-1/3}$ is the modified Bessel function of order $-1/3$, and

$$K_{1/3} = \frac{\pi}{2 \sin \pi/3} (I_{-1/3} - I_{1/3})$$ (15)

$$C_0 = N \int_0^\infty \zeta^{1/3} K_{1/3}(\zeta) e^{-\gamma\zeta} d\zeta$$ (16)

$$n_p = \frac{\pi}{2 \sin \pi/3} \, N\zeta^{2/3} \cdot \left[I_{-1/3} \int_0^{\zeta} \zeta^{1/3} I_{1/3} e^{-\gamma\zeta} \, d\zeta \right.$$

$$\left. - I_{1/3} \int_0^{\zeta} \zeta^{1/3} I_{-1/3} e^{-\gamma\zeta} \, d\zeta \right]. \quad (17)$$

All the terms of (14) vanish at $z = \infty$ ($\zeta = 0$). C_0 is so defined that $n = 0$ also at $z = -\infty$. The remaining constant C is determined by the field-aligned electron and ion current densities (i_i and i_e) in the following way. According to (1), (2), (5), and (6) we have

$$i_i = enu_i$$

$$= -\frac{e}{v_{i0}m_i + v_{e0}m_e} \left[Q - \frac{i}{e} v_{e0}m_e \right] \quad (18)$$

$$i_e = -enu_e$$

$$= \frac{e}{v_{i0}m_i + v_{e0}m_e} \left[Q + \frac{i}{e} v_{i0}m_i \right], \quad (19)$$

where

$$Q = e^{z/H_n} \left[nmg + k(T_i + T_e) \frac{dn}{dz} \right]. \quad (20)$$

Q vanishes at $z = -\infty$ ($\zeta = +\infty$), i.e. at low levels i_i and i_e have a fixed relation to each other, given by the mobility ratio.

At $z = +\infty$ ($\zeta = 0$) the only term in (14) that contributes to Q is the one containing C. If the ratio i_i/i_e has a different value at high levels from that at low levels, Q must have a nonvanishing limit at $\zeta = 0$. This means that the constant C must have a nonvanishing value, and consequently, according to (14) that the height distribution of ionization must be different from the current-free case, where $C = 0$.

The value of the constant C, which determines the structural deviation caused by the currents, can be expressed as follows in terms of the topside and low-level current densities $i_{i\infty}$, $i_{e\infty}$ and i_{i0}, i_{e0}

$$C = -\frac{\sqrt{3}\,(1/3)!}{\pi 2^{2/3}} \frac{v_{e0}m_e}{eg\beta(m_i + m_e)} \cdot \left(i_{i\infty} - i_{e\infty} \cdot \frac{i_{e0}}{i_{e0}} \right)$$

$$= \frac{i_{e\infty}}{3.7 \times 10^{-3}} - \frac{i_{i\infty}}{6.3 \times 10^{-6}}, \quad (21)$$

where the current densities are measured in amp/m^2.

From the properties of the solution (14) it further follows that the condition $n > 0$ everywhere implies

$$C > -0.59 = -\hat{C}. \quad (22)$$

This means that if ions were extracted in the absence of precipitation of electrons, only a certain maximum current density $\hat{i}_i = 3.7 \times 10^{-6}$ amp/m² could be extracted. Similarly, if electrons were extracted in the absence of ion precipitation, there would exist another maximum (negative) current density $\hat{i}_e = -2.2 \times 10^{-3}$ amp/m².

The existence of these limiting currents can be understood as follows. When ions are extracted, most of the charge carriers must come from the ionization processes in the ionosphere (because at the bottom of the ionosphere the current is mainly carried by electrons, and very few ions are supplied). However, the height-integrated production of ion pairs in the ionosphere corresponds to a current density of the order of only a few microamperes/meter². When electrons are extracted, most of them are supplied from below. However, a small fraction of the current (namely $m_e v_{e0}/m_i v_{i0}$) is carried by ions at the bottom of the ionosphere. The ions have to be replaced by ionization processes. Hence $\hat{i}_e = (m_i v_{i0}/m_e v_{e0}) \hat{i}_i$.

4. Discussion

$C/\hat{C}$ is the most natural measure of the strength of the structural disturbance in the ionosphere. We may write (21) in the form

$$\alpha = \frac{C}{\hat{C}} = \frac{i_{e\infty}}{2.2 \times 10^{-3}} - \frac{i_{i\infty}}{3.7 \times 10^{-6}}, \tag{23}$$

where the current densities are measured in amperes/meter². It is then evident that the electron current term can become considerable at topside electron currents of the

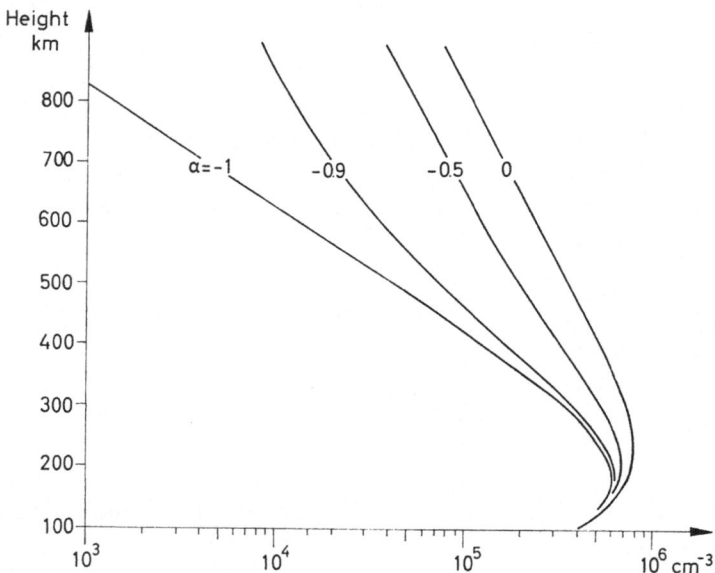

Fig. 2.   Electron concentration distributions calculated from Equation (14) for the values -1, -0.9, -0.5, and 0 of the parameter α. The corresponding values of ion current density in the absence of precipitation are 3.7, 3.3, 1.8, and 0 μamp/m².

order of 10^{-3} amp/m^2, corresponding to a flux of electrons of 10^{16} m^{-2} sec^{-1}. On the other hand, the ion current term can be quite appreciable at a current density of the order of 10^{-6} amp/m^2, corresponding to a flux of 10^{13} ions/m^2 sec.

Figure 2 shows numerically integrated profiles for various negative values of α, which may, for example, correspond to the indicated ion currents in the absence of precipitation. We notice that in addition to the general decrease of ionization density there is also a downward shift of the position of the density maximum. The density decrease at high altitude can be quite substantial. The high densities at low altitudes are due to the simplifying assumption of constant neutral scale height. In nature the scale height decreases at low altitudes. If this had been accounted for in our equations, it would have resulted in more realistic profiles below the F-layer maximum.

Applying the results to the ionosphere we first note that several rocket and satellite observations have indicated currents and particle fluxes large enough to produce α values approaching unity according to (23). Thus Zmuda *et al.* (1966) have observed magnetic disturbances, which have been interpreted by Cummings and Dessler (1967) as due to field-aligned currents of the order of 10^{-6} amp/m^2. Proton fluxes up to 10^{13} protons/m^2 sec for energies below 10 keV appear likely, according to Eather (1967).

These currents must flow during an extended time in order to affect the ionosphere appreciably. The present calculations apply only to stationary states. The time constant for transient phenomena is of the order of three hours. This time can be reduced considerably, if the current density temporarily exceeds the maximum value given above.

We suggest that field-aligned currents be considered as a mechanism, which may at least partly explain F-layer storms, the ionospheric troughs as observed by many investigators (e.g., Muldrew, 1965; Liszka, 1966; Calvert, 1966; Sharp, 1966; Brace *et al.*, 1967; Reddy *et al.*, 1967; Nishida, 1967), and the so-called G and Σ conditions reported by Herzberg and Nelms (1968). The G condition is interpreted as a temporary and localized decrease of electron concentration in the F_2 region, and the Σ condition is interpreted as a very narrow vertical region of electron concentration depletion (minitrough).

If the ionospheric conditions studied in the present model exist locally in some region in the upper ionosphere, there must also be horizontal density gradients at the edge of this region, and the scale height is the same on both sides of the edge, provided $\alpha > -1$. Calvert (1966) has observed such density gradients with the same scale height on both sides. The consequences of diffusion and currents caused by these horizontal gradients have not been included in the present one-dimensional model but should be included in future more complete models.

There are of course also other effects of importance for these phenomena. Thus ionospheric heating probably has a profound effect on the ionosphere, as has been shown by Yonezawa (1963) in his theory of the midlatitude F-layer storm. The currents themselves would lead to heating of the ionosphere and the upper atmosphere that may result in an increased altitude of the F-layer maximum when the total electron content is decreased. The present model predicts a slight downward shift of the position of the F_2 maximum, but the heating probably dominates and causes a net upward shift. This will be further considered in a future paper.

Depletion of the outer magnetosphere by large-scale convection is apparently of great importance outside the plasmapause (Carpenter, 1966). This must also lead to particle losses through an upward plasma convection from the ionosphere. The corresponding reduction of density may be found from our theory by putting $i_{e\infty} = -i_{i\infty} \neq 0$ in Equation (23). Such convection may operate alone or superimposed on finite currents. It may even be more important, but so far no observations exist, which permit a simple estimate of its magnitude.

Kendall and Pickering (1967) have shown that the weak current (10^{-7} amp/m^2) proposed by Dougherty (1963) to flow between conjugate points has little effect on ionospheric density. This is in agreement with our results. However, the currents considered in the present paper are of a different kind and much stronger.

Acknowledgment

The numerical evaluation of the solution for the height distribution by means of a computer was performed by Dr. S. D. Shawhan, whose contribution in this respect is gratefully acknowledged.

References

Alfvén, H.: 1939, 'A Theory of Magnetic Storms and Aurorae, 1', *Kungl. Vet. Akad. Handl.* (3), **18**, No. 3.
Alfvén, H.: 1940, 'A Theory of Magnetic Storms and Aurorae, 2, 3', *Kungl. Vet. Akad. Handl.* (3), **18**, No. 9.
Alfvén, H.: 1950, *Cosmical Electrodynamics*, Oxford University Press, Oxford.
Birkeland, K.: 1913, *The Norwegian Aurora Polaris Expedition 1902–1903*, vol. 1, 2nd ed., Christiania.
Bostrom, R.: 1964, 'A Model of the Auroral Electrojets', *J. Geophys. Res.* **69**, 4983.
Brace, L. H., Reddy, B. M. and Mayr, H. G.: 1967, 'Global Behavior of the Ionosphere at 1000 Kilometers Altitude', *J. Geophys. Res.* **72**, 265.
Calvert, W.: 1966, 'Steep Horizontal Electron-Density Gradients in the Topside *F*-Layer', *J. Geophys. Res.* **71**, 3665.
Carpenter, D. L.: 1966, 'Whistler Studies of the Plasma-pause in the Magnetosphere', *J. Geophys. Res.* **71**, 693.
Cummings, W. D. and Dessler, A. J.: 1967, 'Field-Aligned Currents in the Magnetosphere', *J. Geophys. Res.* **72**, 1007.
Dougherty, J. P.: 1963, 'Some Comments on Dynamo Theory', *J. Geophys. Res.* **68**, 2383.
Eather, R. H.: 1967, 'Auroral Proton Precipitation and Hydrogen Emissions', *Rev. Geophys.* **5**, 207.
Herzberg, L. and Nelms, G. L.: 1968, 'Proton Flare of July 7, 1966, Ionospheric Conditions as deduced from Topside Soundings', to be published.
Kendall, P. C. and Pickering, W. M.: 1967, 'Magnetoplasma Diffusion at F_2-Region Altitudes', *Planetary Space Sci.* **15**, 825.
Liszka, L.: 1966, 'Latitudinal and Diurnal Variations of the Ionospheric Electron Content near the Auroral Zone in Winter', in *Electron Density Profiles in Ionosphere and Exosphere* (ed. by J. Frihagen), North-Holland Publ. Co., Amsterdam, p. 549.
Muldrew, D. B.: 1965, '*F*-Layer Ionization Troughs deduced from Alouette Data', *J. Geophys. Res.* **70**, 2635.
Nishida, A.: 1967, 'Average Structure of Storm-time Change of the Polar Topside Ionosphere at Sunspot Minimum', *J. Geophys. Res.* **72**, 6051.
Reddy, B. M., Brace, L. H., and Findlay, J. A.: 1967, 'The Ionosphere at 640 Kilometers on Quiet and Disturbed Days', *J. Geophys. Res.* **72**, 2709.

Sharp, G. W.: 1966, 'Midlatitude Trough in the Night Ionosphere', *J. Geophys. Res.* **71**, 1345.

Yonezawa, T.: 1959, 'A New Theory of Formation of the F_2-Layer', *J. Atmospheric Terrest. Phys.* **15**, 89.

Yonezawa, T.: 1963, 'The Characteristic Behavior of the F_2-Layer during Severe Magnetic Storms', in *Proc. Intern. Conf. on the Ionosphere, London, July 1962*, Institute of Physics and Physical Society, London.

Zmuda, A. J., Martin, J. H. and Heuring, F. T.: 1966, 'Transverse Magnetic Disturbances at 1100 Kilometers in the Auroral Region', *J. Geophys. Res.* **71**, 5033.

EFFECTS OF ION-NEUTRAL COLLISIONS ON
ION-ACOUSTIC INSTABILITIES IN
THE AURORAL IONOSPHERE*

N. D'ANGELO

European Space Research Institute, Frascati (Rome), Italy

Abstract. The effect of ion-neutral atom (or molecule) collisions on the excitation of ion-acoustic instabilities in the auroral ionosphere is considered. These collisions may play an important role in selecting the wavelength of the perturbations that are most likely to grow at any given height. The results of the calculations are compared with observational evidence obtained from radio star scintillations and radar aurora echoes.

1. Ion-acoustic oscillations have been studied in great detail both theoretically (Bernstein and Kulsrud, 1960; Jackson, 1960; Fried and Gould, 1961) and in laboratory experiments (Alexeef and Neidigh, 1963; Hatta and Sato, 1962; Wong *et al.*, 1964; Andersen *et al.*, 1968). Detection of electrostatic ion waves in the magnetosphere was reported by Scarf *et al.* (1965), and a possible role played by ion waves in the acceleration of auroral electrons was discussed by Swift (1968). In particular, Swift's analysis shows that sufficiently large currents flow along the earth's magnetic field lines, from the magnetosphere down to the ionospheric auroral regions, for ion-acoustic waves to be excited in magnetospheric regions. In addition, Swift's estimates indicate that conditions for growth of ion-acoustic waves often exist at ionospheric heights.

Granted that excitation of these waves can occur, one may ask which particular wavelength in the ion-acoustic spectrum is more likely to grow at any given height or, more specifically, which mechanism may cause the available oscillation energy to concentrate in some particular range of wavelengths? It will be shown here that ion-neutral atom (or molecule) collisions probably play an important role and that they may be indeed the dominant mechanism in this wavelength selection process. After a preliminary analysis of the ion-acoustic instability in a plasma of variable degree of ionization (Section 2), the most unstable wavelength (λ_{crit}) is computed as a function of height (Section 3). The results are then compared with experimental observations available from radar auroral echoes and radio star scintillations, both of which indicate sizes for ionospheric 'irregularities' in fair agreement with the calculations (Section 4). Finally, we speculate on the possible relevance of our calculations to ionospheric hiss and certain types of micropulsations (Section 5). In Section 6 the conclusions are presented.

2. In this section we discuss briefly the excitation of ion-acoustic waves in a homogeneous plasma consisting of positive ions, electrons, and a neutral gas background. The calculations are not given in detail, as the type of analysis involved has been

* This is a reprint of a paper published in *J. Geophys. Res.* **73** (1968) 6313.

commonly used in the past. Only the novel features, which are significant to our present purposes, receive a more detailed consideration.

Wave propagation occurs in a direction parallel to the magnetic field lines. The ion and the electron distribution functions are Maxwellian with temperatures T_i and T_e, respectively. The electron distribution, however, is shifted with respect to the ion distribution so that electrons have some 'drift' velocity, μ, along the field lines. The neutral gas background enters the problem only through a collision term in the Boltzmann equation for ions. Electron-neutral collisions can be neglected. The collision term in the Boltzmann equation for the ions is written as $-v(f-f_0)$, where v is the ion-neutral collision frequency and f and f_0 are the total and the undisturbed ion distributions, respectively. This way of treating ion-neutral collisions generally over-estimates their effect. Better ways of dealing with the same problem have been discussed, for instance, by Bhatnagar *et al.* (1954). We may note in this connection that a more precise treatment than ours would mainly result in some shift toward larger values of λ_{crit} for the curve in Figure 3 of Section 3. It will, however, be apparent from the rest of the discussion that the present analysis is adequate for our purposes.

By linearization of the Boltzmann equations for the ions and the electrons and Poisson's equation with perturbations of the type $e^{i(kx-\omega t)}$ (k real, ω complex), the following dispersion relation is obtained

$$2K^2D^2 = Z^1\left(\omega/kv_{i,\,th} + iv/kv_{i,\,th}\right) + \left(T_i/T_e\right)Z^1\left(-\mu/v_{e,\,th}\right),$$

where $\omega = \omega_r + i\omega_i$ is the wave frequency, $v_{i,\,th}$ and $v_{e,\,th}$ are the ion and electron thermal velocities, respectively, and $D^2 = \kappa T_i/4\pi n e^2$. Z^1 is the derivative of the plasma dispersion function defined, for instance, by Fried and Conte (1961). We introduce also a parameter ε as the ratio D/λ_{i-n} between D and the ion-neutral collision mean free path. By using tables of the plasma dispersion function (Fried and Conte, 1961), one may construct instability diagrams of the type shown in Figure 1 for the case

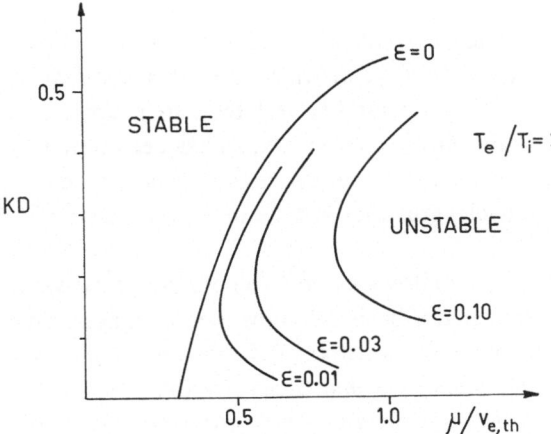

Fig. 1. Instability diagram for ion-acoustic oscillations. The KD, $\mu/v_{e,\,th}$ plane is divided into stable and unstable regions. ($T_e/T_i = 3$, $\varepsilon = 0$, 0.01, 0.03, 0.10.)

$T_e = 3T_i$ and for several values of ε. The value $\varepsilon = 0$ corresponds, of course, to the case of no collisions. The result obtained in this case is the same as Jackson's (1960). As ε is increased, however, one notices a slight increase of the critical value of $\mu/v_{e,\,th}$ required for the instability. In addition, it is no longer the very large wavelengths ($k \approx 0$) that are first excited. Instead, the value of KD corresponding to the minimum value of $\mu/v_{e,\,th}$ for which instability occurs, here called $K_{crit}D$, is now finite and a monotonically increasing function of ε. The presence of the neutral gas acts against the excitation of long wavelengths, a fact physically very reasonable if one keeps in mind that these wavelengths correspond to small values of the wave frequency, for which $\omega_r \lesssim v$. Results similar to those of Figure 1 have also been obtained for the case $T_i = T_e$. They are not given here as they exhibit no different features. In Figure 2 the quantity $K_{crit}D$ is shown as a function of ε, for $T_e/T_i = 3$.

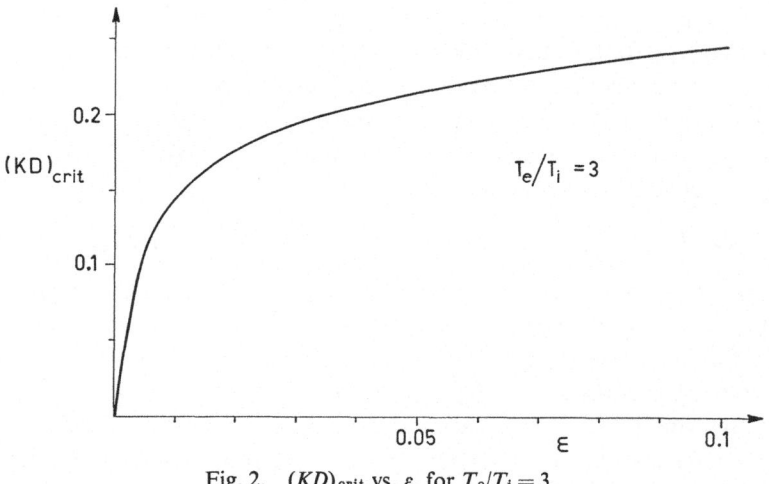

Fig. 2. $(KD)_{crit}$ vs. ε, for $T_e/T_i = 3$.

It seems clear that when current is passed through a plasma one will generally observe wavelengths $\lambda_{crit} = 2\pi/K_{crit}$, provided no other selective mechanism for the wavelength is operative. The instability will tend to prevent a rise of the current through an increase of the effective resistivity, as has been noted by Buneman (1959). A self-regulating mechanism, therefore, assures that generally the observed oscillations will be those having wavelengths that are the most easily excited ($K \approx K_{crit}$).

3. The analysis of Section 2 allows us now to construct a curve of λ_{crit} vs. height in the auroral ionosphere on the basis of known or assumed values for the various ionospheric and atmospheric parameters. This has been done by utilizing data contained in *Handbook of Geophysics and Space Environments* (1965). These data are representative of average ionospheric conditions rather than of the more special situation of the auroral zone, particularly during active auroras. It is felt, however, that for the kind of answer required here, the 'average' data provide a sufficiently good basis. For orientation purposes we may mention here that the ion collision cross sections appropriate to our calculations are of the order 5×10^{-15} cm^2–10^{-14} cm^2, as can be easily found from the data in *Handbook of Geophysics and Space Environments* (1965). They are of the same order of 'resonant' charge-exchange cross sections of atmospheric ions (temperature ~ 0.1 eV) with atoms of the same kind. The electron-atom cross sections are also roughly comparable. However, as already mentioned in Section 2, electron-atom collisions do not play a major role in our computations. It may also be noted at this point that, although the value of μ required for the instability depends strongly on the ratio T_e/T_i, λ_{crit} is entirely insensitive to any variation of this ratio between at least 1 and 3, if $\varepsilon \lesssim 0.1$ ($h \geq 100$ km). Figure 3 shows our computed curve of λ_{crit} vs. height. This curve was obtained by first computing

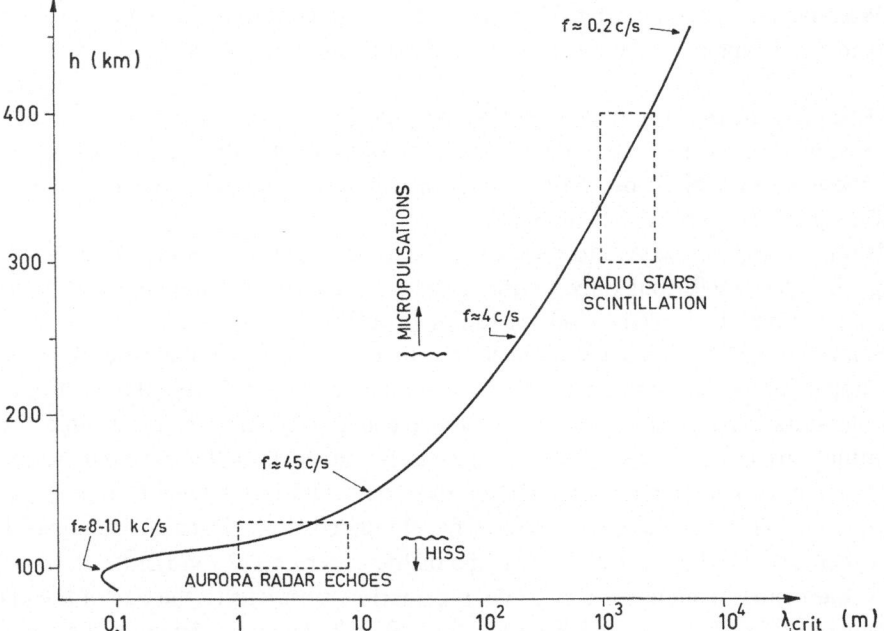

Fig. 3. h(km) versus λ_{crit}(m). Indicated by dotted-line 'boxes' are the locations in the h, λ plane of the ionospheric 'irregularities', as deduced from radio star scintillations and radar auroral echoes.

the quantity ε as a function of height. The curve of ε vs. h, however, is not shown here for the following reason. The quantity ε is about 5×10^{-3} at $h \simeq 105$ km and decreases very rapidly with increasing h. Therefore, inspection of Figure 2 reveals that for heights $h \gtrsim 105$ km, we must utilize only the linear portion of the $(KD)_{crit}$ vs. ε diagram, where $(KD)_{crit} = \text{constant} \times \varepsilon$. This last relation can, of course, be rewritten as $\lambda_{crit} = \text{constant } \lambda_{i-n}$, so that essentially only a knowledge of λ_{i-n} vs. h is necessary for our purposes over most of the height range considered in this paper. It should be kept in mind that according to our remarks in Section 2, a more precise treatment of ion-neutral collisions would shift the entire curve of Figure 3 to somewhat larger values of λ_{crit}. As it will appear from a comparison with experimental observations (Section 4), this shift improves the agreement with the observations. Figure 3 indicates which value of λ is likely to be observed at any given height, provided the excitation conditions are fulfilled. That they are often fulfilled in the auroral ionosphere, however, is inferred by extending the estimates of Swift (1965) for the magnetospheric plasma down to E- and F-region heights. The argument is that used by Swift and will therefore not be repeated here. We may remark, however, that particularly at F-region heights the instability is helped by the ratio T_e/T_i reaching values as large as 3 or 4. This fact, as is well known, reduces the Landau damping acting on the waves. In addition, although detailed measurements are apparently lacking, it seems reasonable to expect that in active auroras T_e/T_i may perhaps reach values even larger than those mentioned above.

We must now examine what kind of observational confirmation of our results exists and to what type of phenomena our calculations may be relevant.

4. Estimates of the size of ionospheric 'irregularities' come from two different types of observations: (a) scintillations of radio stars and (b) radar echoes from auroras. To these we may also add the information gained by observing the scintillations of radio signals from artificial satellites.

It is not our purpose here to review the extensive literature existing on these subjects (e.g., Booker, 1960), but only to quote some of the information we have been able to collect and which seems relevant to our problem.

Scintillations of radio stars are produced by ionospheric 'irregularities' localized in the upper ionosphere, the phenomenon being analogous to the twinkling of optically visible stars. The precise location of the plasma irregularities responsible for the scintillations is not known, but it is generally agreed that the irregularities of importance in this connection are most often to be found in the height range 300 km$\lesssim$ $h\lesssim$400 km. The irregularities are somewhat elongated along the earth's magnetic lines, and their longitudinal size appears to be of the order of 1–2 km (Little *et al.*, 1962).

Radar echoes from auroras provide information mostly as to the size of the plasma irregularities at E-region heights ($h\sim$ 100–130 km). Here the irregularities are also elongated along the geomagnetic lines, and the 'parallel' length is estimated to be of the order of several meters (Leadabrand and Hodges, 1967; Titheridge, 1963; McClure, 1964).

In regard to both scintillations of radio stars (Little *et al.*, 1962) and scintillations of radio signals from satellites (Munro, 1963), it may be noted that the scintillations are much more intense in auroral regions than at temperate latitudes. Little *et al.* (1962) remark that "this great intensity is to a large extent connected with the presence of aurora and its associated terrestrial effects".

In Figure 3 we have indicated by dotted-line 'boxes' the approximate location of the irregularities in the λ, h diagram, as deduced from the above sets of observations. The fit with our computed curve of λ_{crit} vs. height is satisfactory. One may notice the variation of approximately three orders of magnitude in the irregularity size from

E-region to F-region heights, which is predicted by the calculations and actually observed.

At heights between the E-region and the peak of the F-region there seems to be generally a lack of observation of the irregularities (Leadabrand et al., 1965). One may at first wonder whether this is a consequence of the observational technique rather than a true absence. However, Carpenter and Colin (1963), in discussing a correlation between whistler-mode propagation and high-frequency north scatter, explicitly remark that "echoes are not observed from altitudes between the E- and F-regions for reasons that are little understood". In the next section, while no reason is offered for this apparent lack of observation, it is pointed out that it may well be the counterpart of a similar 'gap' in the frequency spectrum of the so-called 'emissions', extending from the 'hiss' frequencies down to the frequency range of geomagnetic micropulsations.

5. The contents of this section are of a *very* speculative nature, and the tentative suggestions advanced are to be taken with great caution. It is felt, however, that a possible association of the irregularity gap, remarked upon in the previous section, with a gap in the frequency spectrum of the so-called 'emissions' is probably worth discussing in some detail. Whatever opinion the reader may form as far as the association between these two 'gaps' is concerned, it will be realized that our conclusions in Sections 2, 3, and 4 are in no way affected.

In Figure 3 we have indicated at several heights the frequency of ion-acoustic oscillations with $\lambda = \lambda_{crit}$. Notice that in the height range $100 \text{ km} \lesssim h \lesssim 120 \text{ km}$, the frequency of these ion-acoustic oscillations is in the range $10 \text{ kc/s} \gtrsim f \gtrsim 0.5 \text{ kc/s}$ (the range of the 'hiss' frequency). Similarly, the height range, $250 \text{ km} \lesssim h \lesssim 500$–$600 \text{ km}$ would correspond to periods, T, between $\sim 0.2 \text{ sec}$ and $\sim 100 \text{ sec}$. A gap in the 'emission' spectrum, between $f \approx 0.5 \text{ kc/s}$ (the lower limit of the 'hiss' spectrum), and $T \sim 0.2 \text{ sec}$ (the lower limit of the micropulsation period spectrum), would then correspond to a gap in the height of the ionospheric 'irregularities' between $h \approx 120 \text{ km}$ and $h \approx 250 \text{ km}$.

The conclusion one would be tempted to draw is that at least some of the 'emissions' do originate in plasma (ion-acoustic) instabilities in certain layers of the ionosphere. We must examine, however, the arguments in favor of this interpretation as well as the arguments against it.

Among the first, we may quote (a) the association sometimes reported between auroral radar echoes with micropulsations, and (b) the evidence offered, among others, by Tepley and Wentworth (1962) and Campbell and Matsushita (1962). To quote from Tepley and Wentworth (1962): "... from recent experimental observations it seems more likely that at least some of these micropulsations are related to particle streams spiraling down along geomagnetic field lines, and to electric currents flowing in the lower ionosphere". Campbell and Matsushita (1962) observe that "all evidence seems to indicate that the micropulsation storm results from an ionospheric current system set up at times of primary electron bombardment". They add that "during an appearance of giant pulsations Harang found that the radio echo reflected from a virtual height of 600 to 800 km showed amplitude pulsations with exactly the same period as the magnetic giant pulsations. Vestine reported an occasion of simultaneous appearance of short-period micropulsations and longitudinal oscillations in luminosity of weak, thin homogeneous auroral arcs". A further point in favor is that ion-acoustic oscillations may produce micropulsations of about the right intensity. A time varying density perturbation must give rise to a perturbation in the local value of the magnetic field. From a pressure balance argument (valid at least for the low-frequency oscillations we are here concerned with) we find that $\Delta B/B \approx \frac{1}{2}\beta\Delta n/n$, β being the ratio between material and magnetic pressure at the location of excitation. For the F peak, with $\beta \approx 10^{-4}$, $B \approx 0.4$ gauss and a $\Delta n/n \approx 0.1$–0.2 we obtain a ΔB of some 200–400 mγ.

On the other hand, there are at least two arguments indicating that, for certain types of 'emissions', the interpretation in terms of ion-acoustic wave excitation in the lower ionosphere is almost certainly untenable. The first argument is connected with the measured polarization of several types of micropulsations (Troitskaya, 1967), which is not easily accounted for in our model. The second is the fact that emissions are often found to exhibit a well-defined frequency (or frequency band). In our model one should rather expect several frequencies being excited simultaneously in the different ionospheric layers, unless special and somewhat ad hoc conditions were postulated favoring excitation in a relatively narrow height range. One such special condition would prevail, for instance, with a ratio T_e/T_i considerably larger than unity only over a fairly restricted height range.

Altogether, it would seem that our suggestion is probably relevant for the case of 'storm' or 'burst-like' micropulsations, which exhibit neither a well-defined frequency nor a well-defined polarization. For the more regular 'emissions' the interpretation usually given in terms of magnetohydrodynamic waves would appear more plausible.

6. The calculations presented in this paper indicate that ion-neutral collisions may act as a powerful selection mechanism for the wavelength of plasma 'irregularities' produced by current flow along the geomagnetic lines at E- and F-region heights in the auroral ionosphere. A comparison of the predicted irregularity sizes at various heights with observational evidence, obtained from radio star scintillations and radar aurora echoes, shows a satisfactory agreement. The frequency of the ion-acoustic oscillations that (given a sufficient electric current along the geomagnetic lines) are excited at E-region heights falls in the range of the so-called 'hiss'. The ion-acoustic frequencies in the F-region are, on the other hand, in the range of geomagnetic micropulsations.

We have considered here the excitation by current flow along the geomagnetic lines of the simplest type of low-frequency ion-oscillations. The analysis is restricted to phenomena occurring in or near the auroral regions. Other types of instabilities are likely to be much more important at low latitudes, particularly in the region of the equatorial electrojet. These are the instability discussed by Farley (1963) and the 'universal' instability predicted by Krall and Rosenbluth (1963) and observed in laboratory plasmas by D'Angelo and Motley (1963), Lashinsky (1964), and Buchelnikova (1966). Whether ion-neutral collisions play a role in selecting the parallel wavelengths in this type of instability may be an interesting problem for further research.

Acknowledgments

Drs. L. Enriques, T. S. Green, and H. L. Jordan read the manuscript and offered very useful criticism.

References

Alexeef, I. and Neidigh, R.: 1963, 'Observations of Ionic Sound Waves in Plasmas: Their Properties and Applications', *Phys. Rev.* **129**, 516.

Andersen, H. K., D'Angelo, N., Jensen, V. O., Michelsen, P. and Nielsen, P.: 1968, 'Effects of Ion-Atom Collisions on the Propagation and Damping of Ion-Acoustic Waves', *Phys. Fluids* (to be published).

Bernstein, I. B. and Kulsrud, R. M.: 1960, 'Ion-Wave Instabilities', *Phys. Fluids* **3**, 937.

Bhatnagar, P. L., Gross, E. P. and Krook, M.: 1954, 'A Model for Collision Processes in Gases', *Phys. Rev.* **94**, 511.

Booker, H. G.: 1960, 'Radar Studies of the Aurora', in *Physics of the Upper Atmosphere* (ed. by J. A. Ratcliffe), Academic Press, New York and London.

Buchelnikova, N. S.: 1966, 'Diffusion Perpendicular to the Magnetic Field in the Universal Instability', *Nucl. Fusion* **6**, 122.

Buneman, O.: 1959, 'Dissipation of Currents in Ionized Media', *Phys. Rev.* **115**, 503.

Campbell, W. H. and Matsushita, S.: 1962, 'Auroral-Zone Geomagnetic Micropulsations with Periods of 5 to 30 Seconds', *J. Geophys. Res.* **67**, 555.

Carpenter, G. B. and Colin, L.: 1963, 'On a Remarkable Correlation between Whistler-Mode Propagation and High-Frequency North Scatter', *J. Geophys. Res.* **68**, 5649.

D'Angelo, N. and Motley, R. W.: 1963, 'Low-Frequency Oscillations in a Potassium Plasma', *Phys. Fluids* **6**, 422.

Farley, D. T., Jr.: 1963, 'Two-Stream Plasma Instability as a Source of Irregularities in the Ionosphere', *Phys. Rev. Letters* **10**, 279.

Fried, B. D. and Conte, S. D.: 1961, *The Plasma Dispersion Function*, Academic Press, New York and London.

Fried, B. D. and Gould, R. W.: 1961, 'Longitudinal Ion Oscillations in a Hot Plasma', *Phys. Fluids* **4**, 139.

Handbook of Geophysics and Space Environments (ed. by S. L. Valley), McGraw-Hill, New York, 1965.

Hatta, Y. and Sato, N.: 1962, 'The Experimental Study of the Ionic Wave in the Dark Plasma', *Proc. Fifth Intern. Conf. Ioniz. Phenom. in Gases, Munich*, North-Holland Publ. Co., Amsterdam.

Jackson, E. A.: 1960, 'Drift Instabilities in a Maxwellian Plasma', *Phys. Fluids* **3**, 786.

Krall, N. A. and Rosenbluth, M. N.: 1963, 'Low-Frequency Stability of Non-Uniform Plasmas', *Phys. Fluids* **6**, 254.

Lashinsky, H.: 1964, 'Universal Instability in a Fully Ionized Inhomogeneous Plasma', *Phys. Rev. Letters* **12**, 121.

Leadabrand, R. L. and Hodges, J. C.: 1967, 'Correlation of Radar Echoes from the Aurora with Satellite-Measured Auroral Particle Precipitation', *J. Geophys. Res.* **72**, 5311.

Leadabrand, R. L., Schlobohm, J. C. and Baron, M. J.: 1965, 'Simultaneous Very High Frequency and Ultra-high Frequency Observations of the Aurora at Fraserburgh, Scotland', *J. Geophys. Res.* **70**, 4235.

Little, C. G., Reid, G. C., Stiltner, E., and Merritt, R. P.: 1962, 'An Experimental Investigation of the Scintillation of Radio Stars observed at Frequencies of 223 and 456 Megacycles per Second from a Location Close to the Auroral Zone', *J. Geophys. Res.* **67**, 1763.

McClure, J. P.: 1964, 'The Height of Scintillation-producing Ionospheric Irregularities in Temperate Latitudes', *J. Geophys. Res.* **69**, 2775.

Munro, G. H.: 1963, 'Scintillation of Radio Signals from Satellites', *J. Geophys. Res.* **68**, 1851.

Scarf, F. L.: 1962, 'Micropulsations and Hydromagnetic Waves in the Exosphere', *J. Geophys. Res.* **67**, 1751.

Scarf, F. L., Crook, G. M., and Fredericks, R. W.: 1965, 'Preliminary Report on Detection of Electrostatic Waves in the Magnetosphere', *J. Geophys. Res.* **70**, 3045.

Swift, D. W.: 1965, 'A Mechanism for Energizing Electrons in the Magnetosphere', *J. Geophys. Res.* **70**, 3061.

Tepley, L. R., and Wentworth, R. C.: 1962, 'Hydromagnetic Emissions, X-Ray Bursts, and Electron Bunches', *J. Geophys. Res.* **67**, 3317.

Titheridge, J. E.: 1963, 'Large-Scale Irregularities in the Ionosphere', *J. Geophys. Res.* **68**, 3399.

Troitskaya, V. A.: 1967, 'Micropulsations and the State of the Magnetosphere', in *Solar-Terrestrial Physics* (ed. by J. W. King and W. S. Newman), Academic Press, New York and London.

Wong, A. Y., Motley, R. W. and D'Angelo, N.: 1964, 'Landau Damping of Ion-Acoustic Waves in Highly Ionized Plasmas', *Phys. Rev.* **133**, A436.

ROLE OF THE UNIVERSAL INSTABILITY IN
AURORAL PHENOMENA*

N. D'ANGELO

European Space Research Institute (ESRIN), Frascati (Rome), Italy

Abstract. The role of the 'universal' instability in auroral phenomena is discussed. Excitation of the instability in the non-homogeneous magnetospheric plasma at $R_E \approx 6$ may modulate particle precipitation into the ionospheric auroral regions and account for some of the observations on pulsating (optical) auroras and X-ray pulsations. It is shown that a good correlation exists between the temporal and spatial variations of particles precipitation and the properties that might reasonably be expected from a drift wave model.

1. Observations of the spatial and temporal character of fast variations in auroral X-rays (Anger *et al.*, 1963; Brown *et al.*, 1965; Barcus *et al.*, 1966; Milton and Oliven, 1967; Parks, 1967; Barcus and Rosenberg, 1965) as well as of pulsating auroras (e.g. Cresswell and Davis, 1966) reveal features strikingly similar to those reported from laboratory experiments on the so-called 'drift' waves (D'Angelo and Motley, 1963; Buchelnikova, 1964; Lashinsky, 1964; D'Angelo *et al.*, 1963; D'Angelo and Von Goeler, 1966; Hendel *et al.*, 1967).

The results of these laboratory experiments have been analyzed in terms of either the so-called 'universal' instability (e.g. Krall and Rosenbluth, 1965) or the 'resistive' instability (e.g. Chen, 1964). The agreement between these theories and the laboratory experiments is good. The 'universal' instability theory of Krall and Rosenbluth (1965) is appropriate to situations in which the ion–ion mean free path is very much larger than the parallel wavelength, $\lambda_\parallel$, whereas the 'resistive' instability theory applies to the opposite case.

We wish here to discuss the relevance of the 'universal' instability theory to the observations of pulsating (optical) auroras and fast quasi-periodic variations of the auroral X-ray fluxes.

The principal result of the present investigation is that a good correlation exists between the temporal and spatial variations of particle precipitation and the properties that might reasonably be expected from a 'drift' wave model.

2. In view of the application to hot and tenuous magnetospheric plasmas we summarize here the results of the collisionless theory of Rosenbluth (1965).

In the collisionless 'drift' instability theory low-frequency modes are involved, with $\omega \ll \omega_c$, ω_c being the ion cyclotron frequency. Suppose we are dealing with a low-β plasma (β being the ratio between particle and magnetic pressure). In this case an *electrostatic* treatment is adequate, in which fluctuations of the magnetic field are

---

* This is a reprint of a paper published in *J. Geophys. Res.* **74** (1969) 909.

neglected and the electric field is derived from a potential. We take the **B** field (steady and uniform) in the positive z-direction of a Cartesian frame of reference and the density gradient ∇n, in the negative x-direction. For the case $T_i = T_e = T$ the ion and electron diamagnetic drift velocities (oppositely directed along the y-axis) are both given by

$$v_d = \left| \frac{1}{n} \frac{dn}{dx} \right| \frac{\kappa T}{M \omega_c} = R_L v_{i,\,th} \left| \frac{1}{n} \frac{dn}{dx} \right|, \tag{1}$$

M being the ion mass, R_L the ion gyro radius and $v_{i,\,th}$ the ion thermal speed.

A linearized theory in which the perturbed quantities vary as $\exp[i(K_y y + K_z z - \omega t)]$ gives the following dispersion relation

$$\omega = K_y v_d \frac{e^{-Z} I_0(Z)}{2 - e^{-Z} I_0(Z)}$$
$$- i \frac{2}{\sqrt{\pi} |K_z| \sqrt{2KT/m_e}} \frac{(K_y v_d)^2}{} \frac{e^{-Z} I_0(Z) |1 - e^{-Z} I_0(Z)|}{\{2 - e^{-Z} I_0(Z)\}^3}, \tag{2}$$

m_e being the electron mass, $Z = K_y^2 R_L^2$ and $I_0(Z) = J_0(iZ)$. The following inequalities are satisfied:

$$K_z^2 \ll K_y^2 \tag{3}$$

and

$$v_{e,\,th} > \omega/K_z > v_{i,\,th}. \tag{4}$$

For $Z = K_y^2 R_L^2 \ll 1$, which is the case in the application of these considerations to the magnetospheric plasma (Section 3), the quantity $e^{-Z} I_0(Z)/[2 - e^{-Z} I_0(Z)]$ is essentially unity. We obtain, therefore, a phase velocity in the y-direction (in the direction of the *electron* diamagnetic drift) equal to v_d of Equation 1.

We may here remark that Rosenbluth's theory (Rosenbluth, 1965) strictly applies to the case of very low-β plasmas. When β is not too low there are modifications to the universal instability theory, which have been discussed by Mikhailovskii and Rudakov (1963). However, even for β's as large as ~ 0.1, the real part of the frequency remains of order $K_y v_d$ (Equation 2) and the inequality $v_{e,\,th} > \omega/k_z > v_{i,\,th}$ (Equation 4) still applies.

In addition Mikhailovskii and Rudakov (1963) remark that the growth rates for the cases of small β and of large β are similar, differing only by a numerical factor.

We therefore seem justified in using the simpler Rosenbluth theory in Section 3. The results of the universal instability theory for the $\beta \ll 1$ case have been used also by Swift (1967) in an attempt to explain the origin of long-period micropulsations.

3. In this section we outline a model for the application of the universal instability theory to the magnetospheric plasma ($R_E \approx 6$) connected along the geomagnetic field lines to the ionospheric auroral regions.

A tenuous, hot plasma is present in the anti-solar magnetosphere at $R_E \approx 6$. We assume that at least part of the time the density of the hot plasma decreases as one moves away from the Earth, and call l the scale height for the radial density profile. 'Drift' waves may then grow in this region and propagate in the direction of $\nabla n \times \mathbf{B}$, i.e. from W to E. The density perturbations will be strongly elongated along the $\mathbf{B}$ lines, while having much shorter dimensions normal to $\mathbf{B}$.

If precipitation of auroral particles occurs along $\mathbf{B}$ down to the ionospheric auroral regions, this precipitation may be modulated by the waves here considered. This is equivalent to the assumption, which seems quite plausible, that the modulation in the precipitated particle flux at any point in the auroral region is roughly proportional to the instantaneous modulation of the plasma density in the magnetosphere (at $R_E \approx 6$) on the same $\mathbf{B}$ line. To avoid any misunderstanding, we emphasize at this point that we do *not* regard the 'drift' waves as being the agent responsible for particle precipitation. The fact that unmodulated precipitation of auroral particles is apparently observed most of the time would indicate that precipitation is not produced by this agent. Our contention is in fact much more limited. Given particle precipitation, we only wish to show that the *modulation* often reported in connection with pulsating auroras and X-ray pulsations can possibly be understood as being due to 'drift' waves in the magnetospheric plasma.

We must now fix the relevant magnetospheric parameters to use in developing our model. Measurements by Frank (1967) indicate that a hot plasma exists, beyond Carpenter's 'knee', extending to at least $R_E \simeq 10$. Although the spatial structure of this plasma is often complicated, the density is generally decreasing away from the Earth. For at least some of the time a value $l \approx 1000$ km for the scale height seems reasonable. For an argument in favour of this choice we may refer also to Brown *et al.* (1965). As far as the plasma 'temperature' is concerned, we may take $T \approx 1$ keV (10^7 K), although almost any value between ~ 1 keV and ≈ 10 keV is apparently equally well defensible. For the magnetic field at $R_E \simeq 6$ we assume $B \approx 200\gamma$, adopting the simple dipole model. The value of the plasma density does *not* enter the expression for the wave velocity (in which only $(1/n)$ (dn/dx) appears), nor that for the (linear) growth rate.

We should notice here that at times (Frank, 1967) the local magnetospheric plasma density gradient reverses sign from $R_E \simeq 6$ to $R_E \simeq 6.5-7$, the last values of R_E, corresponding to the *northern* boundary of the auroral zone. In such cases the expected propagation of particle precipitation would obviously be from E to W, rather than from W to E.

We are now in a position to estimate the velocity of propagation of a drift wave in the magnetosphere at $R_E \simeq 6$. We obtain $v \approx (1/l)$ $(\kappa T)/(qB) \approx 5$ km/sec, with the above set of parameters. If we let T increase from 1 keV to 10 keV, $v \approx 50$ km/sec. By projecting these values from the equatorial $R_E \simeq 6$ region down to ionospheric heights in the auroral region, one obtains as a plausible range for the velocity at ionospheric heights $v(\text{ionospheric}) \approx 0.5-5$ km/sec.

The ratio $K_\parallel / K_\perp = \lambda_\perp / \lambda_\parallel$ between the perpendicular and parallel wavelengths can be roughly estimated from Equation 4 as being $\sim 10^{-2}-10^{-3}$. If for $\lambda_\parallel$ we assume twice the length of a field line at auroral latitudes ($\lambda_\parallel \approx 2 \times 10^5$ km), we obtain a $\lambda_\perp$, at $R_E \simeq 6$ in the equatorial plane, of 200–2000 km. This means, at ionospheric heights, a wavelength $\lambda_\perp$ (ionospheric) of 20–200 km.

An estimate of the growth rate from the imaginary part of Equation 2 gives a τ_{growth} from somewhat less than a minute to several minutes.

To summarize the predictions of our model for the modulation of particle precipitation in the auroral ionosphere, we predict:

(a) 'Patches' of precipitation somewhat elongated in the auroral zone in the N–S direction, and moving along parallels of geomagnetic latitude.

(b) Propagation from W to E at the *southern* boundary of the auroral zone, but possibly opposite at the northern boundary.

(c) Velocity of propagation roughly in the range 500 m/sec to ~5 km/sec.

(d) Wavelength or size of the 'patches' in the range ~20 km to ~200 km.

(e) Occasional presence of harmonics in the period spectra of particle precipitation.

(f) Growth times of the order of minutes.

The relevant auroral observations are summarized in the next section.

4. The auroral observations relevant to our discussion fall into two different categories: (a) pulsating (optical) auroras, and (b) X-ray pulsations and microbursts.

Observations of pulsating auroras have been reported by Cresswell and Davis (1966). Pulsating auroras occur near the equatorward boundary of auroral displays and only rarely are observed in early evening. Their maximum occurrence is confined generally to a few hours following the local geomagnetic midnight. The observed types include thin arcs, shapeless regions several km across which pulsate rapidly with periods as short as 0.2 sec, and fairly large patches 50 km or more which pulsate periodically with periods of about 10 sec. Pulsations are restricted to relatively small field-aligned volumes. The power spectra of the pulsations at times extend from ~1 sec to ~10 sec, with a peak lying near 8 sec. Pulsating patches are observed to progress, while pulsating, along parallels of geomagnetic latitude, always in the direction West to East with a velocity in the range ~500 m/sec to ~1 km/sec. The velocity of propagation seems to be roughly independent of the pulsation period (see Figure 8 of Cresswell and Davis).

X-ray pulsations and microbursts have been studied by a number of investigators (e.g. Barcus *et al.*, 1966; Barcus and Rosenberg, 1965; Barcus and Christensen, 1965; Brown *et al.*, 1965; Milton and Oliven, 1967; Evans, 1963; Parks, 1967). In the work of Barcus *et al.* (1966) the results of balloon flights are reported. The pulsation events occur preferentially near the equatorward side of the auroral zone and almost exclusively in the midnight to dawn sector. The spatial extent of the region of coherent oscillation is 100–200 km and spectral analysis shows quasi-periods of some 5–10 sec. During one flight peaks were detected in the oscillation spectrum at 4.2, 8.3 and 14.3 sec, which "within the uncertainties, might be considered harmonically related". As in the case of pulsating (optical) auroras, the 'patches' of precipitation move along parallels of geomagnetic latitude. Propagation seems to be also here from W to E, the velocity being typically of a few km/sec, although instances of larger velocities may also be common.

The similarities between the observations of Cresswell and Davis (1966) and Barcus *et al.* (1966) are so striking that it seems hard to doubt they are the manifestations of the same phenomenon.

X-ray microbursts are similar to X-ray pulsations, except for two features: (a) the more irregular character of particle precipitation and (b) the occurrence pattern, microbursts being a feature both of the day-side and the night-side ionospheric auroral regions. Generally speaking, the spatial extent of the zone of coherence for particle precipitation and the period spectra are comparable to those of the X-ray pulsations.

Both the observations of Barcus *et al.* (1966) and of Anderson *et al.* (1966) provide the possibility of estimating, at least in a few instances, the time of growth of the X-ray pulsations. From Figures 1a and b of Anderson *et al.* (1966) and Figures 1, 4 and 12 of Barcus *et al.* (1966) one would estimate growth times from a fraction of a minute to several minutes.

It seems clear that many of the features of pulsating auroras and X-ray pulsations are readily explained on the basis of the universal instability theory. Observations confirm all points (a) to (f) predicted at the end of Section 3. An additional point, however, should still be mentioned. Many observations of X-ray pulsations and microbursts do not provide a direct measurement of the W→E velocity of the precipitation 'patches' but only an estimate of the periods involved and of the horizontal extent of the region over which particle precipitation is coherent. One may plot, as

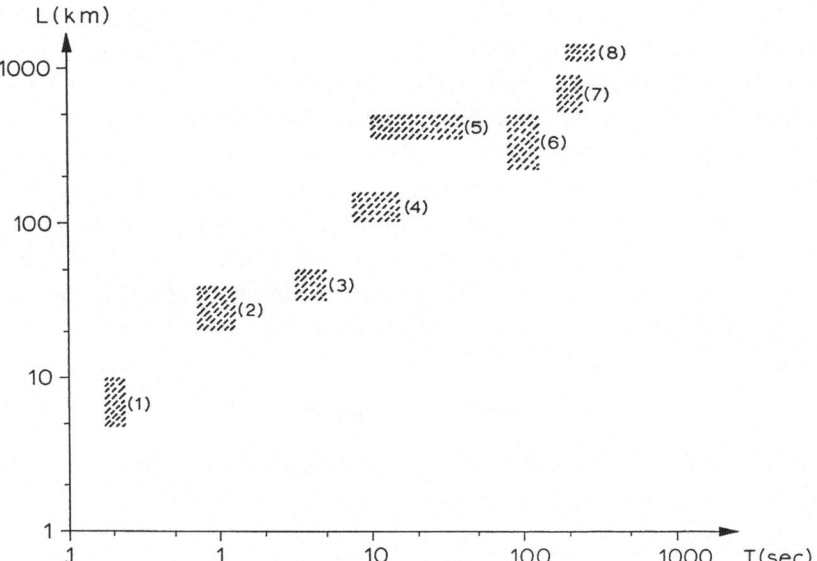

Fig. 1.   Size of the coherent region of particle precipitation, L, vs. the oscillation period; (1) Cresswell and Davis, 1966; (2) Parks, 1967; (3) Cresswell and Davis, 1966; (4) Barcus *et al.*, 1966; (5) Milton and Oliven, 1967; (6) Evans, 1963; (7) Brown *et al.*, 1965; (8) Barcus and Rosenberg, 1965.

we have done in Figure 1, the size of the coherent region, L, vs. the oscillation period, T. Each shaded region in Figure 1 corresponds to the most probable L, T pair out of a single observation. We note that over a three orders of magnitude variation of T, L is a monotonically increasing function of T. The scatter of the shaded regions is large, but it would appear reasonable to consider L proportional to T over the range of values in the diagram. If we make the assumption that the quantity L is equal to $\alpha\lambda_\perp$, $\lambda_\perp$ being the wavelength of the precipitation pattern and α a constant (of order 3–5), we may interpret the diagram in Figure 1 as providing evidence for a *single* propagation process in all cases plotted there, the velocity of propagation being of the order of ~ 1 km/sec.

The fact to be noted is that *all* observations (even those for which v is *not* measured) can be organized in a single diagram, with a value of $\langle v \rangle_{average}$ not far from the value of v obtained in those cases in which a determination of v was possible.

Still two further questions should be discussed here. A survey of the literature indicates that the energy spectrum variations during pulsating events are not the same for the long-period pulsations ($T \sim 1$–10 min), quasi-periodic pulsations ($T \sim 1$–60 sec), and microbursts. In some cases, like the quasi-periodic pulsations of Brown and Weir (1967) the energy spectrum during a pulsation undergoes a systematic variation (hardening as the pulsation develops to peak intensity and softening as it decays). On the other hand modulation on a longer time scale (long-period pulsations) seems to have no large effect on the energy spectrum of the precipitating electrons (Brönstad and Trefall, 1968). One would, therefore, be tempted to argue on this basis against our 'drift' wave interpretation being applicable to the three classes of pulsating events. In fact this argument may be justified were the 'drift' waves regarded as the causative agent for *precipitation*. It does not seem to be justified, on the other hand, when, as we have already made clear, the causative agent of the precipitation is left unspecified and only a *modulation* role is attributed to the drift waves.

A second question to be answered concerns the observations by Brönstad and Trefall (1968), in which propagation is from E to W, rather than from W to E. This may also, at first, seem to argue against the 'drift' wave hypothesis. It must be noted, however, that the observations of Brönstad and Trefall refer to the *northern* part of the auroral zone ($L \approx 6.5$–7) and, as already noted, measurements by Frank (1967) show that at least part of the time the plasma density gradient in the equatorial magnetosphere is directed outward (i.e. away from the Earth) at $L \approx 6.5$–7.

Acknowledgements

I wish to thank Drs. T. S. Green, H. L. Jordan and K. Schindler for reading the manuscript and offering useful comments.

References

Anderson, K. A., Chase, L. M., Hudson, H. S., Lampton, M., Milton, D. W., and Parks, G. K.: 1966, Balloon and Rocket Observations of Auroral-Zone Microbursts', *J. Geophys. Res.* **71**, 4617–4629.

Anger, C. D., Barcus, J. R., Brown, P. R., and Evans, D. S.: 1963, 'Auroral Zone X-Ray Pulsations in the 1- to 15-Second Period Range', *J. Geophys. Res.* **68**, 1023–1030.

Barcus, J. R. and Christensen, A.: 1965, 'A 75-Second Periodicity in Auroral-Zone X-Rays', *J. Geophys. Res.* **70**, 5455–5459.

Barcus, J. R. and Rosenberg, T. J.: 1965, 'Observations on the Spatial Structure of Pulsating Electron Precipitation Accompanying Low-Frequency Hydromagnetic Disturbances in the Auroral Zone', *J. Geophys. Res.* **70**, 1707–1716.

Barcus, J. R., Brown, R. R., and Rosenberg, T. J.: 1966, 'Spatial and Temporal Character of Fast Variations in Auroral-Zone X-Rays', *J. Geophys. Res.* **71**, 125–141.

Brönstad, K. and Trefall, H.: 1968, 'East-West Movements of Pulsating Auroral-Zone X-Ray Events', *J. Atm. Terr. Phys.* **30**, 205–212.

Brown, R. R., Barcus, J. R., and Parsons, N. R.: 1965, 'Balloon Observations of Auroral Zone X-Rays in Conjugate Regions', *J. Geophys. Res.* **70**, 2599–2612.

Brown, R. R. and Weir, R. A.: 1967, 'Energy-Dependent Modulation in Auroral X-Ray Pulsations', *J. Geophys. Res.* **72**, 5531–5536.

Buchelnikova, N. S.: 1964, 'Universal Instability in a Potassium Plasma', *Nucl. Fusion* **4**, 165–168.

Chen, F. F.: 1964, 'Normal Modes for Electrostatic Ion Waves in an Inhomogeneous Plasma', *Phys. Fluids* **7**, 949–955.

Cresswell, G. R. and Davis, T. N.: 1966, 'Observations on Pulsating Auroras', *J. Geophys. Res.* **71**, 3155–3163.

D'Angelo, N., Eckhartt, D., Grieger, G., Guilino, E., and Hashmi, N.: 1963, 'Excitation of Low-Frequency Waves in a Curved Magnetic Field Geometry', *Phys. Rev. Letters* **11**, 525–527.

D'Angelo, N. and Motley, R. W.: 1963, 'Low-Frequency Oscillations in a Potassium Plasma', *Phys. Fluids* **6**, 422–425.

D'Angelo, N. and Von Goeler, S.: 1966, 'Cesium Plasmas in a Magnetic Field with Variable Curvature', *Nucl. Fusion* **6**, 135–143.

Evans, D. S.: 1963, 'A Pulsating Auroral-Zone X-Ray Event in the 100-Second Period Range', *J. Geophys. Res.* **68**, 395–400.

Frank, L. A.: 1967, 'Several Observations of Low-Energy Protons and Electrons in the Earth's Magnetosphere with OGO-3', *J. Geophys. Res.* **72**, 1905–1916.

Hendel, H. W., Coppi, B., Perkins, F., and Politzer, P. A.: 1967, 'Collisional Effects in Plasma-Drift-Wave Experiments and Interpretation', *Phys. Rev. Letters* **18**, 439–442.

Krall, N. A. and Rosenbluth, M. N.: 1965, 'Universal Instability in Complex Field Geometries', *Phys. Fluids* **8**, 1488–1503.

Lashinsky, H.: 1964, 'Universal Instability in a Fully Ionized Inhomogeneous Plasma', *Phys. Rev. Letters* **12**, 121–123.

Mikhailovskii, A. B. and Rudakov, L. I.: 1963, 'The Stability of a Spatially Inhomogeneous Plasma in a Magnetic Field', *JETP* **17**, 621–625.

Milton, D. W. and Oliven, M. N.: 1967, 'Simultaneous Satellite and Balloon Observations of the Same Auroral-Zone Precipitation Event', *J. Geophys. Res.* **72**, 5357–5361.

Parks, G. K.: 1967, 'Spatial Characteristics of Auroral-Zone X-Ray Microbursts', *J. Geophys. Res.* **72**, 215–226.

Rosenbluth, M. N.: 1965, 'Microinstabilities in Plasma Physics', Int. Atomic Energy Agency (IAEA), Vienna 1965.

Swift, D. W.: 1967, 'A New Interpretation of Long-Period Micropulsations', *J. Geophys. Res.* **72**, 4885–4898.

WHISTLERS-USE FOR DETERMINATION OF
COMPOSITION AND TEMPERATURE

STANLEY D. SHAWHAN

Division of Plasma Physics, Royal Institute of Technology, Stockholm, Sweden

Abstract. A review is given of the theory of the 'Eckersley' whistler, the nose whistler, and the ion cyclotron whistler. The technique for deriving the magnetospheric electron concentration from nose whistlers and the results concerning the 'knee' in the density profile and the plasmapause are discussed. From ion cyclotron whistlers the ion concentration, the electron concentration, the ion gyrofrequencies, and the ion temperatures in the ionosphere can be obtained. The method is described and representative results are presented. The use of other whistler mode noise such as the lower hybrid resonance noise for the determination of composition and temperature is also discussed.

1. Introduction – Theory of Whistlers

Whistlers are characterized as VLF (very-low-frequency) radio noise phenomena whose source is external to the ionosphere and whose dispersion characteristics in frequency and time are determined by the plasma properties of the ionosphere and magnetosphere along the path of energy propagation [1]. The naturally occurring whistlers have as their source the electromagnetic energy radiated by a lightning discharge in the atmosphere. The classical 'Eckersley' whistler and its high latitude form, the nose whistler, obtain their dispersion characteristics from the interaction of the wave with the electrons along the path in the magnetosphere. This energy may be reflected back and forth between hemispheres along the terrestrial magnetic field lines to produce multiple traces observable in both hemispheres. Ion cyclotron whistlers, however, result from the wave interaction with the heavy ions in the ionosphere and in general cannot be observed in the magnetosphere or on the ground. The theory of these whistler types is briefly reviewed.

1.1. 'ECKERSLEY' WHISTLERS

'Eckersley' whistlers are whistlers which have a frequency-time dispersion characteristic that fits the empirically derived Eckersley dispersion law [2]

$$t = Df^{-1/2},$$ (1)

where D is a constant for a given path that depends on the variation of the magnetic field strength and the electron density along the path. These whistlers appear as a descending frequency on a frequency-time spectrogram and may obey this relation within a frequency range from a hundred hertz to several kilohertz. These whistlers are observed between geographic latitudes of $\pm 80°$ but are rare at high latitudes.

Storey [3] explained this whistler phenomenon as a dispersed form of the energy from a lightning discharge which has propagated along the magnetic line of force from one hemisphere to the other. He verified Equation (1) theoretically. Figure 1 shows how

D'Angelo (ed.), Low-Frequency Waves and Irregularities in the Ionosphere. All rights reserved

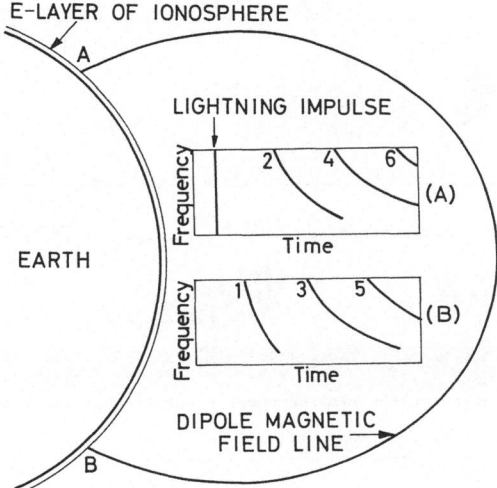

Fig. 1.   Field-line path followed by Eckersley and nose whistlers [1].
(With permission from Stanford University Press.)

whistler trains can occur at conjugate points A and B. Energy from a lightning source enters the ionosphere at point A. Because the lightning discharge emits all frequencies at a given instant in time the initial signal (arrow) appears as a vertical line on a frequency-time spectrogram. Lower frequency components travel more slowly because of the interaction with the electrons along the path. At B some of the energy is reflected in the E-region and some leaks below the ionosphere and appears as trace 1 described by Equation (1). The reflected energy appears at point A as trace 2 but $D_2 = 2D$. As the energy reflects back and forth traces 3, 4, etc. may be observed with dispersions $3D$, $4D$, etc.

1.2. Nose Whistlers

At high latitudes the observed whistlers do not necessarily obey Equation (1) and in fact there appears a frequency for which the time delay is a minimum. This frequency of minimum delay is called the nose frequency and the whistler is called a nose whistler. Several examples of nose whistlers are shown in Figure 2 with nose frequencies that range from 1.2 to 4.4 kHz. For latitudes where the maximum frequency observed is nearly equal to the minimum gyrofrequency along the path, the time delay for the wave frequency components is given more exactly by

$$t = \frac{1}{2c} \int\limits_{\text{path}} \frac{\pi_e (2\omega - \Omega_e)}{\omega^{1/2} (\omega - \Omega_e)^{3/2}} \, ds, \tag{2}$$

where ω = wave frequency; π_e = electron plasma frequency = $(4\pi Ne^2/m)^{1/2}$; N = electron number density; Ω_e = electron gyrofrequency = eB_0/mc; e = electronic charge; m = electronic mass; B_0 = magnetic field magnitude.

The term $\omega^{1/2}(\omega - \Omega_e)^{3/2}$ in the denominator of Equation (2) shows that the time

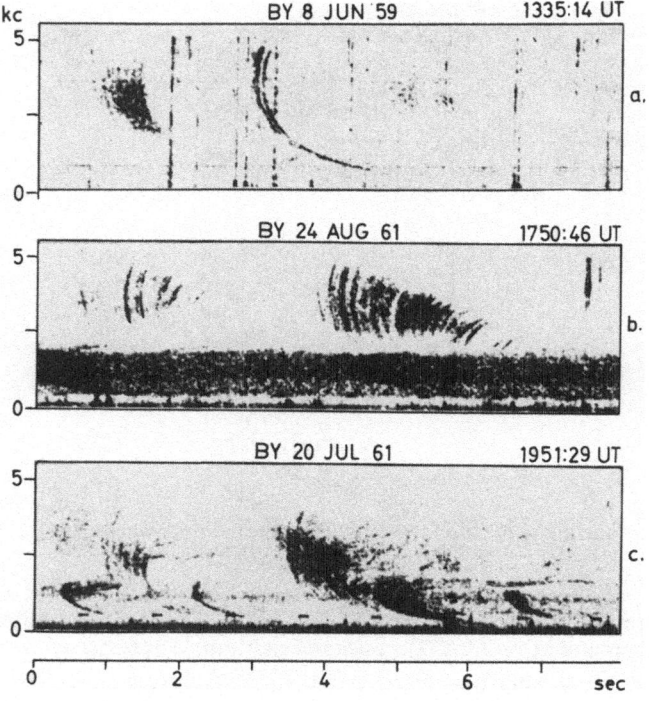

Fig. 2.   Examples of high latitude nose whistlers [1].
(With permission from Stanford University Press.)

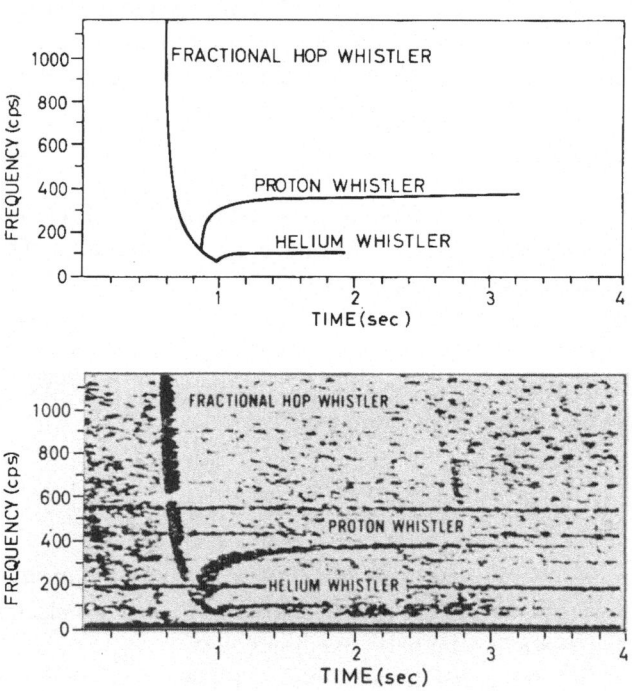

Fig. 3.   Spectrogram of proton and helium ion cyclotron whistlers [5].
(With permission from the North-Holland Publishing Company.)

delay integral should increase for ω small and for ω approaching the electron gyro-frequency Ω_e. Therefore a nose frequency must occur and its value depends on the variation of the electron density N and magnetic field B_0 along the path.

1.3. Ion cyclotron whistlers

At frequencies below 1 kHz whistler traces similar to those shown in Figure 3 have been observed in satellite data [4, 5]. Simultaneously with a descending whistler frequency there may exist one or more frequencies which initially rise and approach a constant frequency lasting for several seconds. These additional whistlers have been explained has being a dispersed form of the original lightning energy and the dispersion is due to the effects of ions along the propagation path [6].

For a plasma that contains electrons and several ions two different whistler modes, one right-hand and one left-hand polarized, can propagate for certain frequency ranges and altitudes which depend on the concentrations and the magnetic field strength. In Figure 4B the important frequencies which affect these two modes are plotted against altitude for the model ionosphere (electrons, H^+, He^+, O^+) given in Figure 4A. The curves marked Ω_1 and Ω_2 are the proton and helium gyrofrequencies, $D=0$ is called the crossover frequency and $L=0$ the cutoff frequency. At the gyrofrequency wave energy for the left-hand (L) polarized mode is absorbed. The cutoff frequency

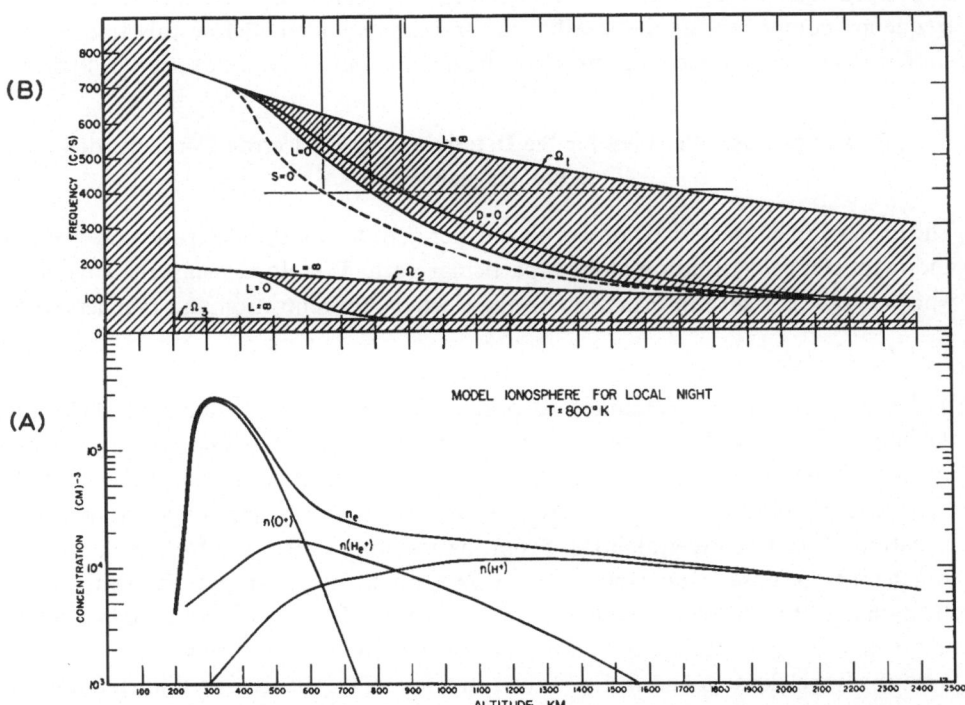

Fig. 4. Diagram showing variation of frequencies related to the origin of ion cyclotron whistlers as a function of attitude [6]. (With permission from Stanford University Press.)

is the frequency at which the L wave is reflected. At the crossover frequency the wave polarization may change from left-hand polarized to right-hand (R) or vice-versa.

Consider a wave frequency of 400 Hz emitted by the lightning discharge and propagating upward in the ionosphere. If the wave is left-hand polarized it cannot propagate to 800 km at all (see Figure 4). The R wave is not affected at the altitude for $L=0$, but at $D=0$ the wave polarization changes and it becomes an L wave. At 1700 km this L wave is absorbed because the wave frequency is equal to the proton gyrofrequency. At this absorption the time delay for the wave becomes infinite. Consider now all the frequency components that can reach 1700 km. Frequencies above 400 Hz are absorbed at lower altitudes. Frequencies below 400 Hz have suffered a polarization change at altitudes closer to the satellite and travel faster to the satellite for lower frequencies. A frequency component of about 200 Hz changes polarization at the satellite altitude.

These facts then explain the proton whistler trace presented in Figure 3. Wave energy propagates from the lightning source to the satellite. Above the maximum proton gyrofrequency the wave reaches the satellite as a fractional hop 'Eckersley' whistler – an R wave. The component equal to the proton gyrofrequency at the satellite changes from an R to an L wave and, because of the energy absorption, takes a very long time to reach the satellite, which explains the nearly constant tone for several seconds. Lower frequencies couple some energy to the L wave at $D=0$ and appear both as earlier R waves and as increasing earlier L waves. For the crossover frequency component at the satellite the traces are joined. Below this frequency a similar phenomenon occurs to produce the helium whistler trace as seen in Figure 3.

2. Use of Nose Whistlers for the Determination of Electron Concentration in the Magnetosphere

The dispersion characteristic of a nose whistler depends on the electron concentration and magnetic field variation along a particular path. Therefore, measurements of the whistler trace can give information on the electron concentration and magnetic field.

2.1. Technique

If the whistler energy is assumed to propagate nearly along a magnetic field line, the integral over ds in Equation (2) is over a dipole magnetic field line from a particular initial latitude. Then the variation for the magnetic field B_0 is also known. If the electron concentration is assumed to vary in proportion to the magnetic field then Equation (2) can be integrated [1]. From the resulting expression for time delay it is found that the nose frequency occurs at 39% of the *minimum* gyrofrequency along the path and that the major contributions to the time delay come from the ionosphere and from the region at the top of the field line. The time delay through the ionosphere is relatively well known and can be subtracted for all frequencies so that the resulting time-delay–frequency profile of an observed nose whistler depends on the particular field line of propagation and on the electron concentration near the equatorial plane.

A measurement of the nose frequency, then, gives the minimum gyrofrequency along the field line path which is the equatorial gyrofrequency. From this equatorial gyrofrequency the equatorial radius for this path and the magnetic field is specified. The time delay from the time of the lightning discharge depends on the average electron concentration in the equatorial region. For frequencies lower than the nose frequency the whistler dispersion is described by Equation (1). Scaling of a nose whistler trace (below 2.5 kHz in Figure 2a) can give the relation between frequency and time which can be extrapolated to infinite frequency to give the time of the lightning discharge. From this time delay the average equatorial electron density can be obtained.

2.2. RESULTS

At a ground or satellite observing station nose whistlers with different nose frequencies and delay times can be observed. Because the nose frequencies are different the whistlers must have traveled on different field lines. The time delay for different nose whistlers can then give the average electron concentration at different distances from

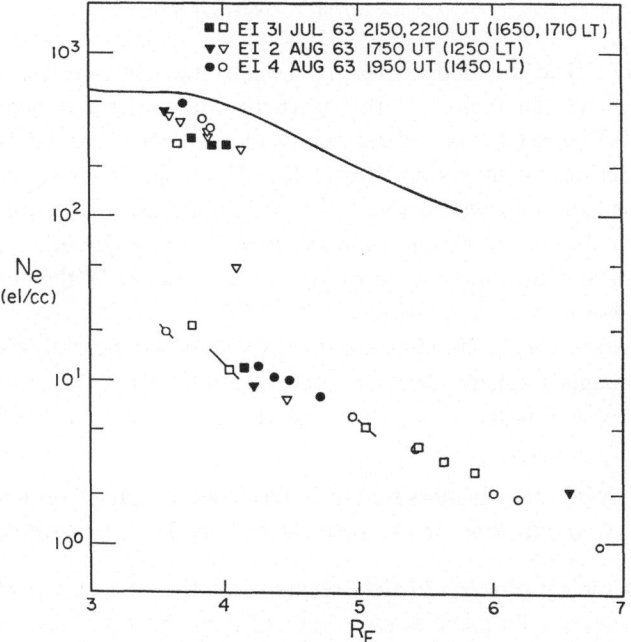

Fig. 5.   Profile of electron concentration vs. geocentric distance in the equatorial plane deduced from nose whistlers [7]. (Reprinted from *J. Geophys. Res.*)

the earth in the equatorial plane. A plot of electron density N_e vs. geocentric distance in earth radii R_E is shown in Figure 5 for $3R_E$ to $7R_E$ [7]. From $3R_E$ to $4R_E$ the electron concentration is several hundred electrons cm^{-3}, but at about $4R_E$ the concentration drops by over a factor of 10 to tens of electrons cm^{-3} and the profile then decreases

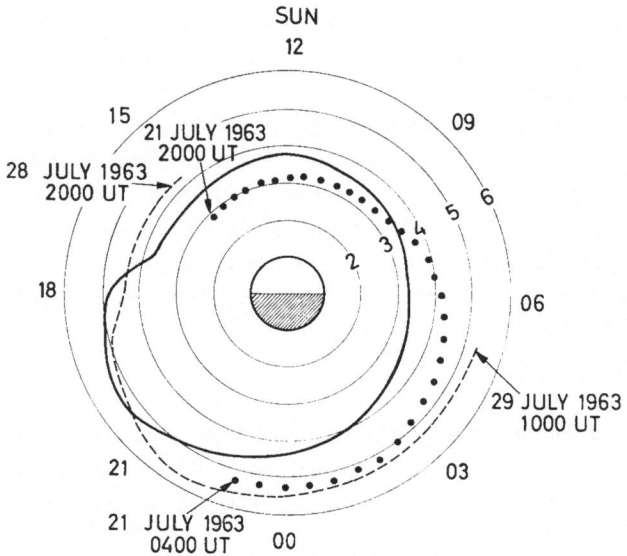

Fig. 6. Equatorial radius of the knee in the electron density as a function of local time [8].
(Reprinted from *J. Geophys. Res.*)

smoothly to $7R_E$. The abrupt decrease in the electron density has been called the 'knee'
by Carpenter, and the region of the magnetosphere where it occurs the plasmapause [7, 8]. In Figure 6 the equatorial radius of this knee is plotted against local time
for different periods of magnetic activity [8]. The solid line shows the position for
moderate but steady magnetic activity, the dotted line for increasing activity and the
dashed line for decreasing activity. During local evening (2000) the knee is farthest
from the earth and during local morning (0600) nearest. With increased magnetic
activity the curve moves closer to the earth.

This nose whistler technique has provided the first and possibly the best measurements of the magnetospheric electron concentration and its variations in the equatorial region. Also this technique does not require the use of satellites or rockets.

3. Use of Ion Cyclotron Whistlers for the Determination of Ion Concentration, Electron Concentration, Ion Gyrofrequency, and Ion Temperature

The theory of ion cyclotron whistlers seems to fit the observed phenomena so well
that the characteristic frequencies and shape of these whistlers can be used to obtain
information about the plasma parameters in the region of the satellite or rocket.
In Figure 7 is shown another spectrogram and a tracing of a proton whistler with
the crossover frequency and proton gyrofrequency indicated. From the ratio of the
crossover frequency to the proton gyrofrequency the fractional concentration of
hydrogen ions can be obtained [9]. The dispersion of the proton whistler trace near
the gyrofrequency can be used to obtain the proton concentration, the electron
density and the proton gyrofrequency [10]. From the abrupt amplitude decrease at

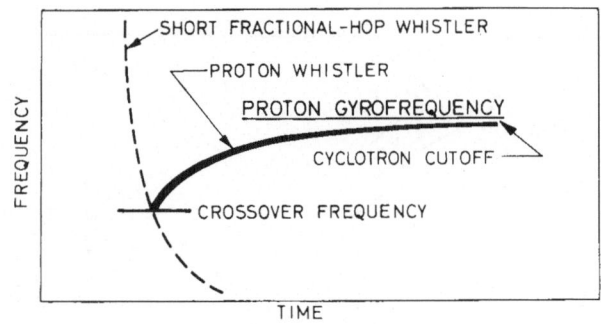

INJUN III   MARCH 5, 1963;   09:12:18 UT

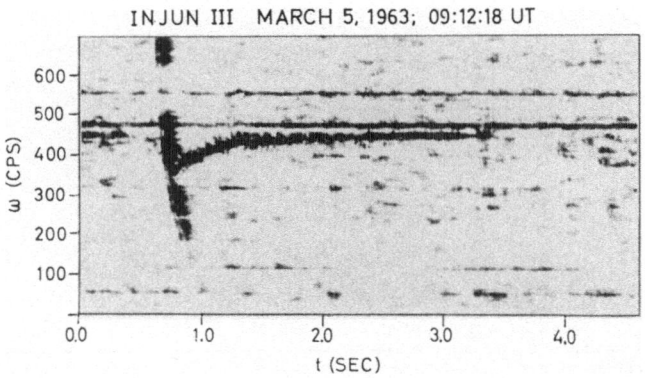

Fig. 7.   Sonagraph and tracing of a proton whistler received by Injun-3 indicating the crossover frequency, proton gyrofrequency and cyclotron cutoff [11].

the end of the proton whistler trace (cyclotron cutoff) the ion temperature can also be obtained [11].

3.1. FRACTIONAL ION CONCENTRATION

As discussed in Section 1.3 the crossover frequency is given by the relation $D=0$ where D is a function of the local magnetic field, the electron density and the fractional concentrations of the ions (H^+, He^+ and O^+ assumed). This condition can be solved to give the fractional concentration of H^+ in terms of the crossover frequency – proton gyrofrequency ratio

$$\frac{n(H^+)}{N_e} = \frac{264}{255}(1 - \Lambda^2),$$ (3)

where $\Lambda = \omega(\text{cutoff})/\Omega_{H^+}$. This formula is accurate to $\pm 3\frac{1}{2}\%$ independent of the other ionic concentrations. Figure 8 shows the fraction concentration (α_1) of H^+ (in %) plotted against invariant latitude and altitude for proton whistlers observed during local summer day (1100–1700) and during local winter night (2300–0500). Two distinct trends in the concentration of H^+ can be seen: (1) the value for α_1 is lower for summer daytime than for winter nighttime at all latitudes and altitudes observed, and (2) the value of α_1 tends to decrease with increasing latitude for a given altitude.

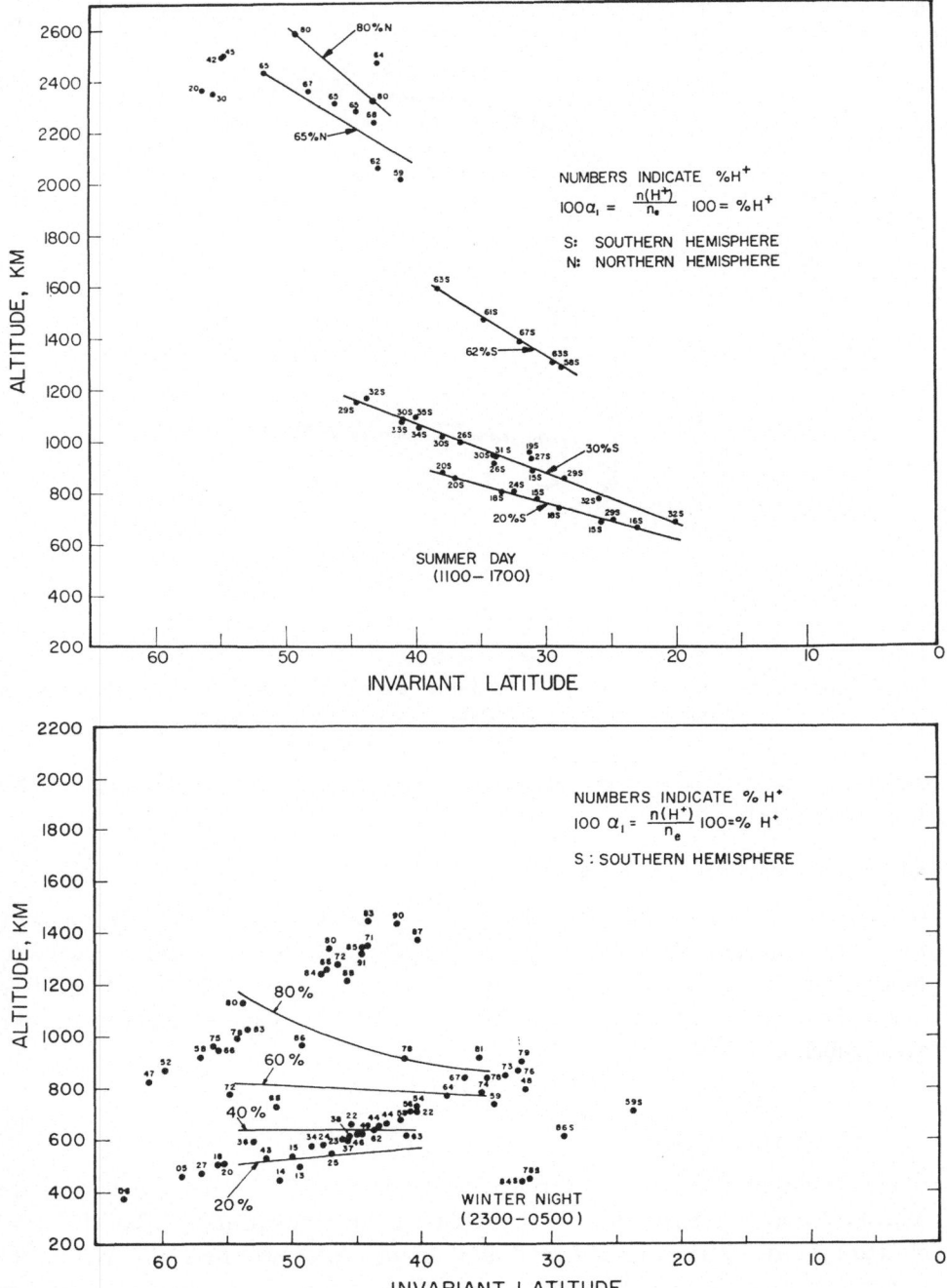

Fig. 8. Plots of the average percentage of H⁺ for winter nighttime and summer daytime from the measurements of proton whistlers [9]. (Reprinted from *J. Geophys. Res.*)

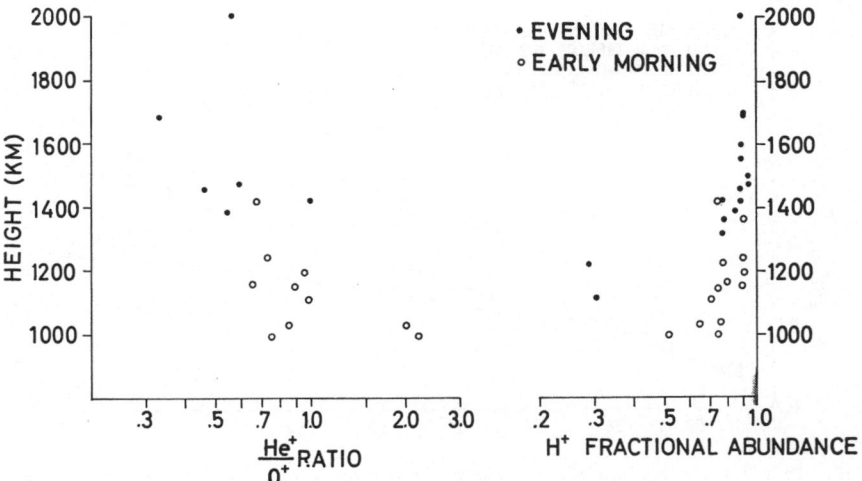

Fig. 9. Variation with height of the fractional concentration of H⁺ and the ratio of He⁺ to O⁺ [5].

These two conclusions are consistent with the view that the heavier ions (He⁺ and O⁺) tend to predominate at a given altitude for higher latitudes, presumably owing to the increased temperature near the auroral latitudes. The diurnal and seasonal temperature variation explains the H⁺ concentration variation.

This technique of obtaining fractional ion concentration has been extended to include helium ion whistlers [5] such as those observed in Figure 3. The He⁺ fractional concentration is plotted in Figure 9 for altitudes of 1000 km to 2000 km. With increasing altitude the H⁺ concentration approaches 100% and the He⁺–O⁺ ratio becomes small.

3.2 Ion concentration, electron concentration and ion gyrofrequency

It can be shown that the time delay of proton whistlers for frequencies very close to the proton gyrofrequency is given by

$$t(\omega) = \frac{\pi_1 \Omega_1^{1/2}}{c\Omega_1'} (\Delta\omega)^{-1/2}, \tag{4}$$

where π_1, Ω_1 Ω_1' and $\Delta\omega$ are the proton plasma frequency, the proton gyrofrequency, the gradient of the gyrofrequency along the path and $(\Omega_1 - \omega)$ respectively. Therefore a plot of $t(\omega)$ vs. $(\Delta\omega)^{-1/2}$ should give a straight line, the slope of which contains the proton density through the proton plasma frequency. A plot of $t(\omega)$ vs. $(\Delta\omega)^{-1/2}$ for different values of the proton gyrofrequency Ω_1^* is shown in Figure 10 for an actual proton whistler. It is assumed that the gyrofrequency for the best straight line fit $(\Omega_1^* = 528.3)$ is the actual gyrofrequency in the region of the satellite. The slope of this line then gives the proton concentration $n(H^+)$. Using the fractional concentration of H⁺ as determined from Equation (3) the electron concentration can also be calculated.

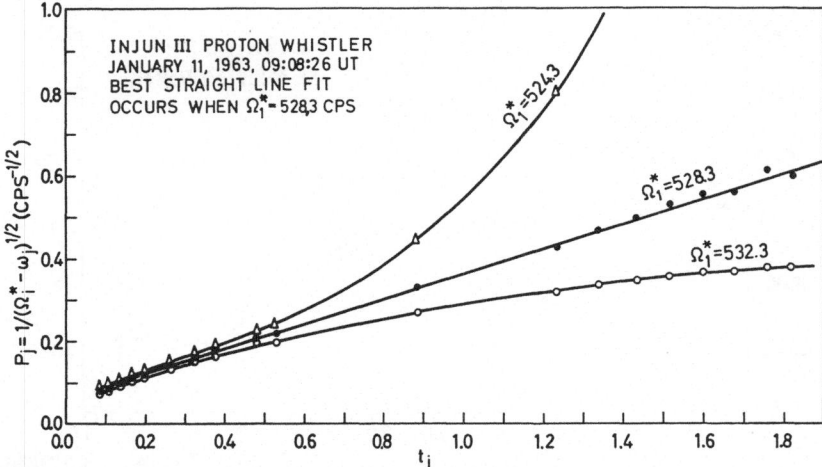

Fig. 10.   Plot showing method of determining the proton density and proton gyrofrequency from the dispersion of proton whistlers [10]. (With permission from Stanford University Press and *J. Geophys. Res.*)

Values of the proton gyrofrequency Ω_1, the proton concentration $n(H^+)$ and the electron concentration n_e are given in Table I for several measured proton whistlers at different altitudes and invariant latitudes. The uncertainty in these values is due to random measurement errors. Values of the proton gyrofrequency calculated from the Jensen and Cain value for the magnetic field are listed and show agreement to better than 1% in all cases. Electron concentration values are also compared to those from the nearest Alouette-1 results at 1000 km with good agreement. This technique can also be used for helium and other ion cyclotron whistlers.

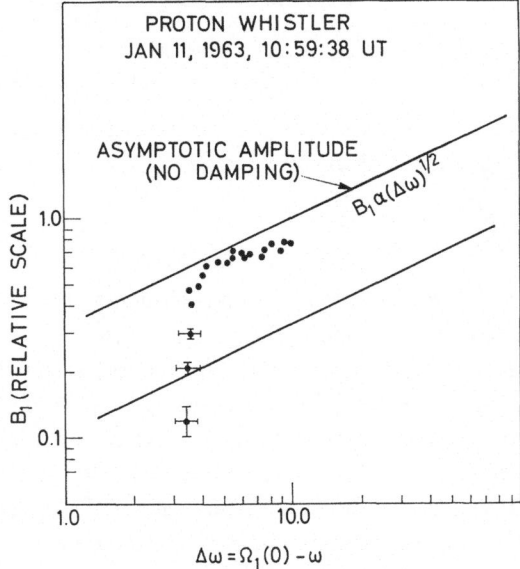

Fig. 11.   The amplitude of the magnetic wave field as a function of the difference between the wave frequency and the proton gyrofrequency showing the abrupt amplitude cutoff [11].

TABLE I (After *J. Geophys. Res.*)

Values of proton gyrofrequency, proton concentration and electron density from measured proton whistlers

Satellite coordinates WN = winter night; SN = summer night					Proton whistler determinations			Nearest Alouette electron density data at 1000 km
UT 1963		Altitude (km)	Invariant latitude (deg)	Jensen and Cain value for $\Omega_1(0)$ (cps)	$\Omega_1(0)$ (cps)	$n(H^+) \times 10^{-3}$ (cm^{-3})	$n_e \times 10^{-3}$ (cm^{-3})	$n_e \times 10^{-3}$ (cm^{-3})
Jan. 7 1008:19	WN	1595	40.2	329.2	329.9 ± 0.5	7.93 ± 0.85	9.51 ± 1.00	
Jan. 7 1013:22	WN	1249	50.2	450.0	451.5 ± 0.8	7.72 ± 1.50	13.5 ± 2.5	7.5–12.5
Jan. 11 0908:26	WN	968	55.8	528.2	528.0 ± 0.9	1.90 ± 0.28	2.97 ± 0.50	3.0–7.5
Jan. 11 1059:38	WN	1308	42.4	408.5	412.3 ± 0.7	14.0 ± 2.6	20.4 ± 3.8	10.0–20.0
Jan. 11 1103:12	WN	1058	51.7	516.4	517.9 ± 1.1	5.94 ± 2.58	7.46 ± 3.21	7.5–12.5
Mar. 3 0746:06	WN	1708	48.9	369.0	373.0 ± 0.6	4.80 ± 0.96	5.93 ± 1.05	
June 11 0142:54	SN	1241	41.1	382.0	382.4 ± 0.3	5.46 ± 0.77	9.80 ± 1.50	7.5–12.5
June 11 0147:06	SN	1535	52.1	387.3	387.5 ± 0.6	6.09 ± 1.13	8.01 ± 1.48	

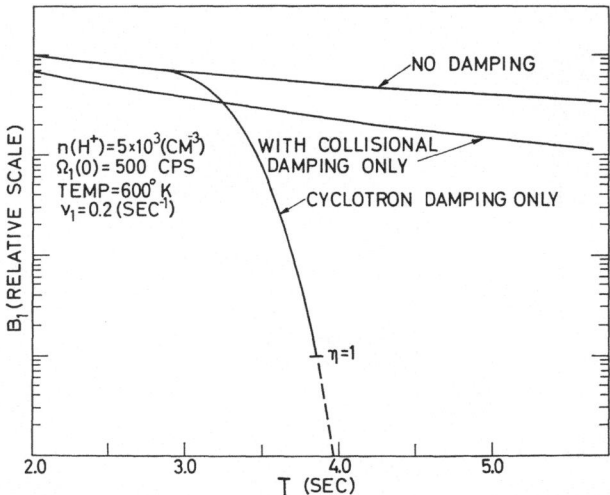

Fig. 12.   Calculated magnetic wave field amplitude as a function of the proton whistler travel time showing that only cyclotron damping gives an abrupt amplitude cutoff [11].

3.3. ION TEMPERATURE

On the proton whistler spectrogram shown in Figure 7 the proton whistler trace seems to disappear abruptly after 3 sec. Figure 11 is a plot of the relative magnetic amplitude of the proton whistler as a function of $\Delta\omega$, the difference between the wave frequency and the proton gyrofrequency. This amplitude shows a very abrupt decrease at $\Delta\omega = 3.5$ Hz. A plot of the expected amplitude decrease with time for the cases of no damping, collisional damping and cyclotron damping is shown in Figure 12. Only the cyclotron damping causes the very abrupt amplitude decrease.

For cyclotron damping of proton whistlers the proton temperature is given by [11]

$$T = \frac{m_1 c^2}{2k\eta_c^3}\left(\frac{\Omega_1}{\Pi_1}\right)^2\left(\frac{\Delta\omega_c}{\Omega_1}\right)^3, \tag{5}$$

where k is Boltzmann's constant and $\Delta\omega_c$ is the value of $\Delta\omega$ at the cutoff. The parameter η_c is determined from the expression

$$\eta_c^{3/2}\exp\left(\eta_c^3\right) = \frac{\pi^{1/2}\cdot\Pi_1\cdot\Omega_1}{\beta\cdot3\cdot\Omega_1'}\left(\frac{\Delta\omega_c}{\Omega_1}\right)^{1/2}, \tag{6}$$

where β is the attenuation at cutoff. Values of π_1 and Ω_1 can be obtained by the method described in Section 3.2. Proton temperatures determined by this method (with correction for the wave normal not being exactly along the magnetic field line) are given in Table II. These temperature values range from 729 K to 1050 K and seem to be lower than those measured by the incoherent radar backscatter technique. Recently two suggestions have been made to improve the ion cyclotron whistler technique [12]. By using a better expansion in the derivation of Equations (5) and (6) and by taking into account the Doppler shift due to the satellite velocity relative to the wave velocity, the calculated temperatures are higher but possibly too high.

TABLE II (After *J. Geophys. Res.*)

Proton temperatures from measured proton whistlers

Data source	Universal time		Altitude (km)	Invariant latitude (deg)	local time (hr)	$\Omega_1(0)$ (cps)	$n(H^+) \times 10^3$ cm^{-3}	Estimated wave normal angle (deg)	Proton temperature (K)
Injun-3	Jan. 7, 1963 10:08:19	WN[a]	1595	40.2	1.71	328.1 ±0.6	7.22 ±0.83	37.6	1050 ±110
Injun-3	Jan. 11, 1963 09:08:26	WN	968	55.8	3.16	527.3 ±1.0	1.77 ±0.62	20.5	746 ±84
Injun-3	Jan. 11, 1963 10:59:38	WN	1308	42.4	1.84	410.0 ±0.6	11.75 ±2.85	34.8	596 ±70
Injun-3	Mar. 3, 1963 07:46:06	WN	1708	48.9	1.52	374 ±0.7	5.37 ±1.09	31.1	684 ±214
Injun-3	June 11, 1963 01:42:54	SN[b]	1241	41.1	2.01	382.3 ±1.0	5.46 ±0.86	36.0	825 ±127
Injun-3	June 11, 1963 01:47:06	SN	1535	52.1	2.40	387.5 ±0.7	6.10 ±1.13	25.2	751 ±179
Alouette-1	Oct. 5, 1964 04:06:02	SN	1000	42.0	23.0	462.7 ±0.6	2.22 ±0.82	33.2	729 ±103

[a] WN = winter night.

[b] SN = summer night

Certainly these techniques of deriving concentration and temperature information can be used for ion whistlers due to the presence of heavier positive ions. It has been suggested [13] that if negative ion whistlers exist they can be used to give the negative ion species, concentration and collision frequency.

4. Use of Other Whistler Mode Phenomena for the Determination of Composition and Temperature

Other naturally occurring noise phenomena can propagate in the electron or ion cyclotron mode and the characteristics of these noises can also be used to obtain information on concentration and temperature. For instance a noise band commonly observed with VLF experiments that measure the electric field component in the ionosphere has a lower cutoff frequency which is very nearly the lower hybrid reso-

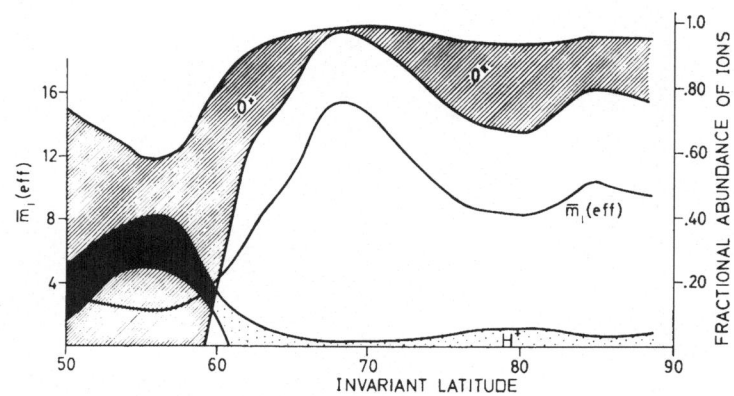

Fig. 13.   Variation of effective ion mass and the resulting ranges of possible O^+ and H^+ concentration values from lower hybrid frequency noise measurements [5].

nance frequency at the experiment location. This lower hybrid resonance frequency depends on the local electron concentration, magnetic field magnitude and relative concentrations of the heavy ions. Therefore from measurements of the lower cutoff frequency of the noise band and the electron concentration information on the relative ion concentrations can be obtained. In Figure 13 the range of possible fractional concentrations of O^+ and H^+ is plotted against invariant latitude from lower hybrid noise and electron density measurements made by Alouette-1 [5]. It is observed that above 60° the ionosphere is predominantly O^+ at 1000 km. Another noise band is often observed with a lower cutoff frequency that seems to be the $L=0$ cutoff frequency [14]. Similar use can be made of this noise band.

Measurements of whistlers and whistler mode noise can be combined to give all of the plasma parameters. As an example proton whistlers can give the proton temperature and the proton and electron concentration and electron density needed to derive the helium and oxygen concentrations from the lower hybrid resonance or $L=0$ noise.

These whistler techniques for the determination of composition and temperature are very desirable because the sounding by naturally occurring radio noise and reception by a passive experiment minimizes the perturbation to the plasma. Of course the sounding occurs at random times and places and neither the sounding signals nor the experiments cover all latitudes, altitudes and local times of interest.

References

[1] Helliwell, R. A.: 1965, *Whistlers and Related Ionospheric Phenomena*, Stanford University Press, Stanford, Calif.

[2] Eckersley, T. L.: 1935, 'Musical Atmospherics', *Nature* **135**, 104–105.

[3] Storey, L. R. O.: 1953, 'An Investigation of Whistling Atmospherics', *Phil. Trans. Roy. Soc. (London)* A**246**, 113–141.

[4] Shawhan, S. D.: 1966, 'Experimental Observation of Proton Whistlers from Injun-3 VLF Data', *J. Geophys. Res.* **71**, 29–45.

[5] Barrington, R. E. and McEwen, D. J.: 1967, 'Ion Composition from VLF Phenomena observed by Alouette-I and -II', *Space Res.* VII, North-Holland Publ. Co., Amsterdam, pp. 624–633.

[6] Gurnett, D. A., Shawhan, S. D., Brice, N. M., and Smith, R. L.: 1965, 'Ion Cyclotron Whistlers', *J. Geophys. Res.* **70**, 1665–1687.

[7] Angerami, J. J. and Carpenter, D. L.: 1966, 'Whistler Studies of the Plasmapause in the Magnetosphere. 2. Electron Density and Total Tube Electron Content near the Knee in the Magnetospheric Ionization', *J. Geophys. Res.* **71**, 711–726.

[8] Carpenter, D. L.: 1966, 'Whistler Studies of the Plasmapause in the Magnetosphere. 1. Temporal Variations in the Position of the Knee and Some Evidence on Plasma Motions near the Knee', *J. Geophys. Res.* **71**, 693–710.

[9] Shawhan, S. D. and Gurnett, D. A.: 1966, 'Fractional Concentration of Hydrogen Ions in the Ionosphere from VLF Proton Whistler Measurement', *J. Geophys. Res.* **71**, 47–59.

[10] Gurnett, D. A. and Shawhan, S. D.: 1966, 'Determination of Hydrogen Ion Concentration, Electron Density, and Proton Gyrofrequency from the Dispersion of Proton Whistlers', *J. Geophys. Res.* **71**, 741–754.

[11] Gurnett, D. A. and Brice, N. M.: 1966, 'Ion Temperature in the Ionosphere obtained from Cyclotron Damping of Proton Whistlers', *J. Geophys. Res.* **71**, 3639–3652.

[12] Quemda, D., Velut, P. M., and Vigneron, J.: 1968, 'Determination of Ionic Temperature from the Propagation Time of Proton Whistlers', NATO Advanced Study Institute on Plasma Waves in Space and in the Laboratory, Røros, Norway, 17–26 April 1968, *Proceedings* (ed. by J. Thomas), University of Edinburgh Press, Edinburgh.

[13] Shawhan, S. D.: 1966, 'Negative Ion Detection in the Ionosphere from Effects on ELF Waves', *J. Geophys. Res.* **71**, 5585–5598.

[14] Gurnett, D. A. and Burns, T. B.: 1968, 'The Low-Frequency Cutoff of ELF Emissions', Univ. of Iowa Res. Rpt. 68–28, Department of Physics and Astronomy, Iowa City, Iowa, 1968; submitted to *J. Geophys. Res.* (1968).

OBSERVATIONS OF
IONOSPHERIC VERY-LOW-FREQUENCY RADIO NOISE*

STANLEY D. SHAWHAN

Royal Institute of Technology, Division of Plasma Physics, Stockholm, Sweden

Abstract. A summary is presented of the AC electric and magnetic fields observed in the ionosphere by the Javelin 8.45 UI sounding rocket launched from Wallops Island, Virginia, on 21 September 1967. The electric dipole and magnetic loop antenna systems and the VLF (30 Hz–70 kHz) receivers are described. Data on the spectral characteristics, the noise amplitude, and the wave field geometry for two types of noise are summarized. A high frequency electromagnetic noiseband occurred between 7 kHz and 30 kHz throughout the flight. Marked spectral changes with altitude, a large ratio of electric to magnetic field amplitude and a field geometry which indicates that the noise is propagating perpendicular to the geomagnetic field were observed. It is argued that the lower cutoff frequency may be the lower hybrid resonance frequency. Precession modulated intense noise bursts were observed on the electric antennas only. This electrostatic noise occurred below 1 kHz in frequency and below 500 km in altitude both upgoing and downgoing. From the spectral shape, the phase of the noise amplitude with respect to the component of velocity perpendicular to the geomagnetic field and the relative electric field amplitudes, it appears that this noise may be caused by a turbulent wake of the payload.

* Full paper: Shawhan, S. D. and Gurnett, D. A.: 1968, 'Recent Rocket Measurements of AC Electric and Magnetic Fields in the Ionosphere', NATO Advanced Study Institute on Plasma Waves in Space and in the Laboratory, Røros, Norway, 17–26 April 1968, *Proceedings* (ed. by J. Thomas), University of Edinburgh Press, Edinburgh.

VLF AND LF EMISSIONS AT AURORAL LATITUDES

T. STOCKFLET JØRGENSEN

Danish Meteorological Institute, Copenhagen, Denmark

1. Introduction

Naturally occurring electromagnetic noise at very low and low frequencies generated in the ionosphere or the magnetosphere is called VLF and LF emission. There are different types of noise classified according to their shape in a frequency–time spectrogram (Gallet, 1959), but they can all be characterized as either discrete noise, which has a clear structure in the spectrogram, or hiss, which has no or very little structure and so resembles white noise.

Satellite observations (Jørgensen, 1968) have shown that the discrete noise as well as band-limited hiss below approximately 2 kHz mainly occur on the day side of the earth equatorwards of the region of auroral activity, whereas wideband hiss above approximately 1 kHz is observed in the auroral region of the ionosphere.

We shall here be dealing exclusively with the wideband hiss which often is called auroral hiss or polar hiss due to its observed correlation with aurora (Martin *et al.*, 1960; Jørgensen and Ungstrup 1962; Morozumi, 1962; Flint, 1968).

2. Observations of Auroral Hiss

An amplitude-vs.-time recording of a typical auroral hiss event as it is often observed on the ground at high latitudes is shown in Figure 1. While the lower cutoff frequency of an auroral hiss burst is of the order of 1 kHz, the upper cutoff frequency is not known. Hiss is often observed at 500 kHz at Byrd Station located in the southern auroral zone, but there may be hiss at still higher frequencies. The event shown in Figure 1 depicts clearly the wideband character of auroral hiss.

The spectral density, as a function of frequency and time, of an ordinary hiss burst in the beginning of an event also observed at Byrd Station is shown in Figure 2. It is seen that the maximum energy is received near 10 kHz.

As the spectrum may be drastically changed when the noise propagates from the magnetosphere to the surface of the earth owing to absorption and/or internal reflection in the ionosphere, it is important to know the noise spectrum observed in space. Such information has been obtained by the VLF and LF receivers in the polar-orbiting OGO-2 satellite, and is reported by Jørgensen (1968). The hiss spectrum observed by the satellite seems to have a maximum near 10 kHz similar to hiss observed on the ground. A typical spectral density at this frequency is 10^{-15}–10^{-14} W m^{-2} Hz^{-1}. Only in one case of 26 noise region passes studied did the spectral density exceed 10^{-12} W m^{-2} Hz^{-1}.

The author (Jørgensen, 1966a) studied the general shape and location of the hiss

D'Angelo (ed.), Low-Frequency Waves and Irregularities in the Ionosphere. All rights reserved

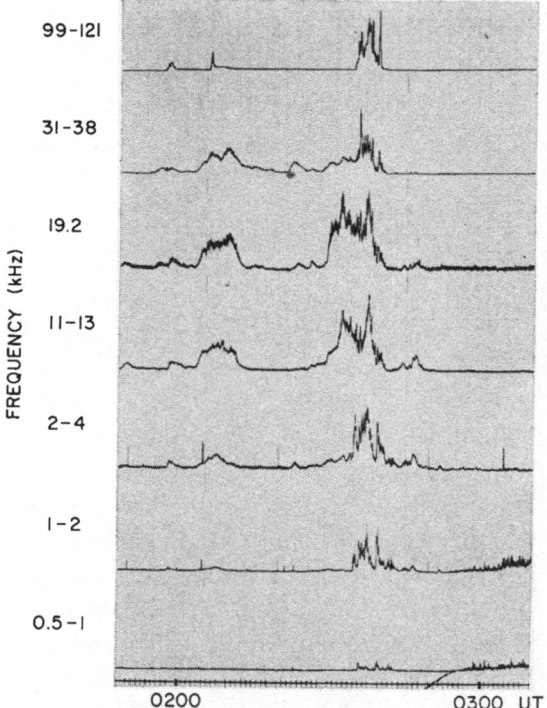

Fig. 1. Auroral hiss, Byrd Station, July 8, 1965. Amplitude (arbitrary units) as function of frequency and universal time.

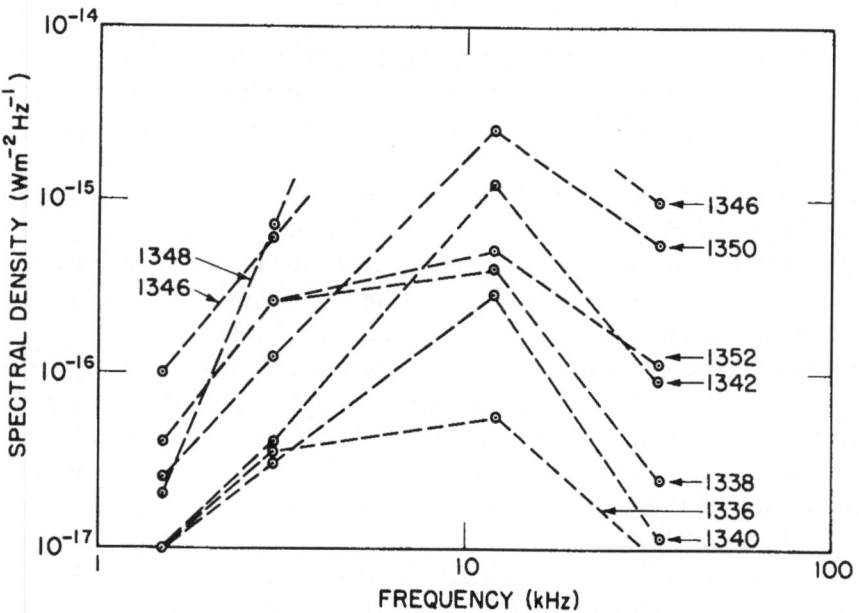

Fig. 2. Auroral hiss, Byrd Station, August 3, 1966. Spectral density as function of frequency and universal time.

zones which are defined as the regions on earth illuminated by hiss propagating down through the ionosphere. The hiss zones – there is one in each of the two polar regions – are fixed not in a coordinate system rotating with the earth, but in an invariant magnetic latitude – local magnetic time coordinate system.

The hiss zones are found to be apparently the same as the zones of auroral precipitation. So they are approximately circularly shaped with their centres situated close to the magnetic midnight meridian some few degrees from the geomagnetic poles. Hiss is not observed equally often everywhere in these zones, as there is a maximum of the occurrence at about 70° invariant latitude shortly before magnetic midnight (Figure 3).

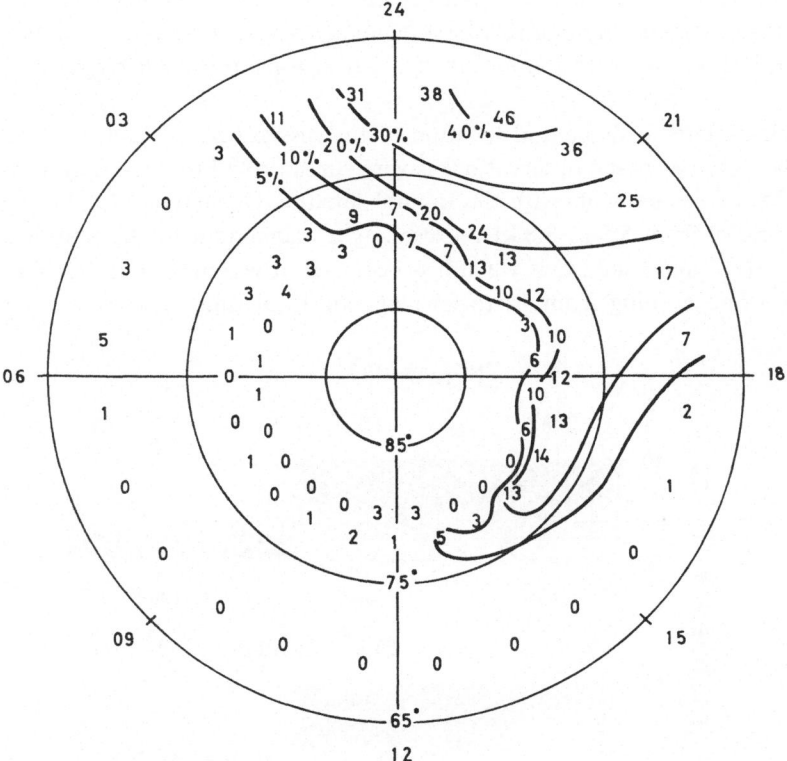

Fig. 3.   Contour map of the 8 kHz hiss zone in the fall (Aug., Sept., Oct.) 1964 drawn in an invariant latitude – local magnetic time coordinate system. The map is based on observations from 3 stations in Greenland. The contours surround regions in which hiss bursts with spectral densities above 10^{-15} W m^{-2} Hz^{-1} occurred in a given percentage of the hourly intervals.

While hiss at about 70° latitude mainly is observed in the evening and until shortly after magnetic midnight, the occurrence of hiss around 80° latitude is somewhat more evenly distributed, and it is possible at this latitude to observe hiss at any time of the day.

Very little investigation of hiss has been carried out close to the magnetic poles, but the observations made so far seem to show that hiss is a relatively rare phenomenon at the very highest latitudes.

The normal low latitude boundary of the hiss zone on the night side of the earth is located around 65° invariant magnetic latitude, consistent with the normal low latitude boundary of the auroral oval.

It is shown (Jørgensen, 1966b) that the general shape of the hiss zone does not change in any pronounced way during the year. The maximum activity on the surface of the earth seems to be higher in the winter than in the summer, which possibly is due to the higher ionospheric absorption in the summer.

During quiet and disturbed conditions the extent of the hiss zone, and the hiss activity as well, is very different. During quiet conditions the hiss activity is small, and hiss is observed only in a relatively small area close to magnetic midnight, while both the hiss activity and the extent of the hiss zone are much bigger in disturbed periods.

The characteristic shape and location of the hiss zone as found from the above-mentioned ground based observations is very similar to a hiss zone found by Gurnett (1966) from measurements with the Injun-3 satellite. Gurnett studied the occurrence of radio noise from 5.5 to 8.8 kHz exceeding a magnetic field strength of 3×10^{-10} gamma^2 Hz^{-1} and found that when hiss occurred, it was almost always during local afternoon and evening from 12 to 24 magnetic local time. The range of invariant

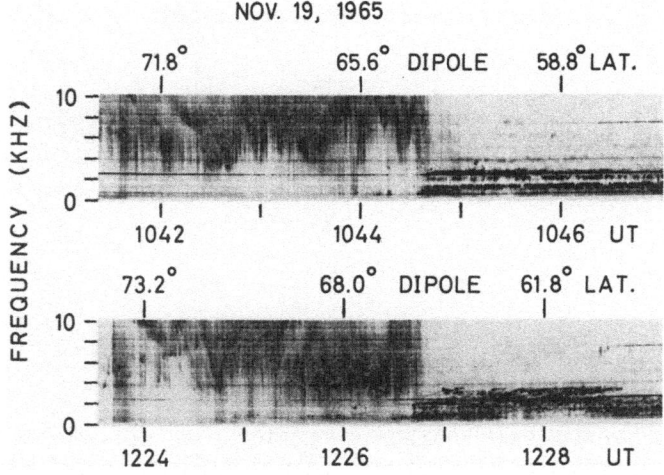

Fig. 4.   VLF spectra of wideband hiss (left) and chorus below 3–4 kHz (right) recorded by OGO-2 on two consecutive passes through auroral regions. In both cases the altitude of the satellite was approximately 600 km and the magnetic local time near midnight.

latitude in which hiss typically occurred was about 10° wide and centered on 77° at magnetic local noon, decreasing to 72° at midnight.

In Figure 4 two VLF spectra recorded by OGO-2 on two consecutive passes through the midnight section of the auroral oval are shown. These recordings are remarkable.

The two noise spectra suddenly change character from wideband auroral hiss with an upper cutoff frequency greater than 10 kHz to what probably is chorus with an upper cutoff frequency at 3–4 kHz. It is very meaningful that the noise spectra recorded during the two consecutive passes are almost identical since in this case it is possible to resolve the usual problem which one faces in connection with satellite measurements, namely whether a change in a measured quantity is due to a change in time or a change of the satellite's position. It is clear in this instance that the sudden change from hiss to chorus happens because the satellite is moving from a region with hiss to a region with chorus, and that these noise regions are quite stationary in space during a period of time at least as long as one orbit period, which for OGO-2 is 104 min. It is remarkable that the boundary between the two regions with hiss and with chorus is so sharp. This indicates that certain physical conditions in the magnetosphere change very abruptly from one region to the other, so the generated noise and perhaps also the mechanism of noise generation is totally changed.

It is difficult to say what this boundary means, but it is interesting to note that it is located at a latitude where the outer trapping boundary and the low latitude boundary of the auroral oval often are found, and where the transition between the open and closed field lines is thought to take place. It is not impossible that this VLF boundary could represent the boundary between the closed magnetosphere and the magnetospheric tail.

At the first passage of the hiss-chorus boundary 1044-45 UT the latitude of the boundary is about 63° dipole latitude (64° invariant latitude), and at the other passage 1226–27 about 66° (67° invariant latitude). At the times of both passes the 3-hourly Kp index was 3−, and according to Starkov and Fel'dshteyn (1967) the position of the low latitude boundary of the auroral oval near magnetic midnight then should be 63°–64° invariant latitude.

In the foregoing a case was described in which the hiss was almost stationary in space during a one-orbit period, but noise patterns which changed very little on three or more consecutive passes have also been recorded by OGO-2 at auroral latitudes.

In Figure 5 another recording of auroral hiss near magnetic midnight is shown. There is here no chorus on the low latitude side of the hiss, but from about 1122:40 whistlers are observed. At approximately 1123:20 an echoing whistler is recorded. The order of events in Figure 5 is very typical: hiss is observed at latitudes where the.

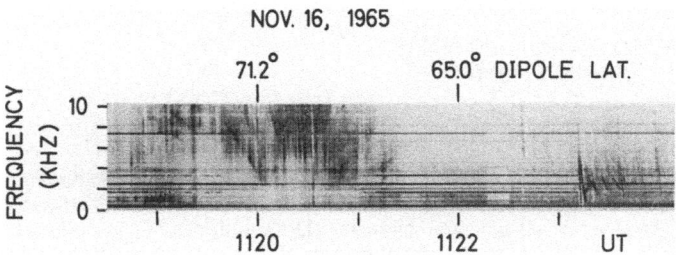

Fig. 5. VLF spectrum showing wideband hiss (left) and whistlers (right) recorded by OGO-2. The altitude of the satellite was approximately 500 km and the magnetic local time near midnight.

auroral oval is expected to be located. The hiss then stops quite suddenly and no emissions except perhaps occasional chorus are observed for 1–2 min as the satellite traverses a latitude interval of the order of 4°–8°. Just as there seems to be a latitudinal boundary below which no hiss is observed, there seems to exist another boundary at lower latitudes above which no whistlers are observed. The whistler boundary is often found at latitudes which make it tempting to correlate it with the plasmapause (Carpenter, 1966). The low latitude hiss boundary in Figure 5 was observed at a time of extremely low geomagnetic activity with Kp=0o at about 68° dipole latitude, which also is the expected latitude of the low latitude boundary of the auroral oval at very quiet conditions (Feldstein, 1966).

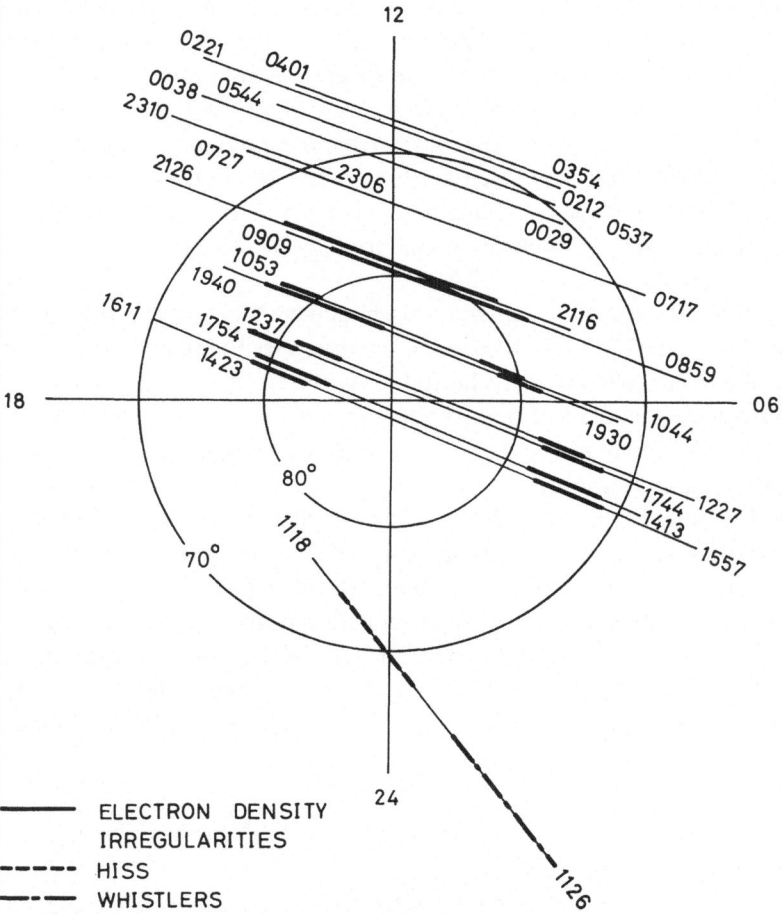

Fig. 6. Fourteen Explorer-20 passes and one OGO-2 pass in the North-polar ionosphere on Nov. 16, 1965, plotted in a magnetic (dipole) latitude–magnetic time coordinate system. The OGO-2 pass is the one crossing 70° at magnetic midnight. The length of the passes indicate the periods of data collection. Time of beginning and end of periods of data collection is given in UT. The observed electron density irregularities and the observation of hiss are located where the auroral oval is expected to be.

Recently Jørgensen and Bell (1968) studied the relation between electron density irregularities and auroral hiss by means of electron density data from the Explorer-20 topside sounder and VLF data recorded by OGO-2 in order to obtain information on the propagation conditions for auroral hiss. In Figure 6 the results of 14 Explorer-20 passes and one OGO-2 pass over the Northern polar region during a 20-hour period are shown. It is seen that on 6 Explorer-20 passes on the dayside of the earth below about 75° dipole latitude no irregularities were found, whereas irregularities were observed on the 8 passes closest to the geomagnetic pole. On the 2 passes tangent to the 80° parallel the satellite traversed one wide irregular region, but on the 6 passes at still higher latitudes two irregular regions were observed on each pass. The irregular regions are grouped quite nicely in a U-shape, and it is attractive to think that the satellite here has observed the morning, noon and afternoon parts of the auroral oval. A comparison of the position of the observed U-shaped region of electron density irregularities, which were observed on an extremely quiet day with 3-hourly Kp indices being 0o, 1−, 0+, 0o, 0o, 0+, 0o and 0+, with the position of the auroral oval at Kp=0–1 as determined by Feldstein (1966), shows quite good agreement. Complete agreement cannot be expected, because Feldstein used corrected (invariant) magnetic latitudes and Jørgensen and Bell (1968) used dipole magnetic latitudes.

Shown schematically in Figure 6 are the VLF phenomena observed on a pass by OGO-2 on Nov. 16, 1965. (The hiss record for this pass is shown in Figure 5.) It is seen that the location of the observed hiss at 67°–74° dipole latitude (68°–75° invariant latitude) fits well with the expected position of the auroral oval, which for Kp=0–1 is located at 67°–72° invariant latitude (Feldstein, 1966).

Even with the lack of direct evidence of temporal and spatial correlation between the hiss and the electron density irregularities, the data are clear enough to suggest strongly that hiss occurs in the regions of electron density irregularities because (1) both phenomena are always observed at locations where the auroral oval is expected to be, (2) the sizes of the areas occupied by hiss and irregularities are of the same order, (3) the small scale structures in the hiss spectra and in the topside soundings of the irregularities look very similar, (4) hiss is observed to occur in regions of auroral particle precipitation (Gurnett, 1966), (5) the electron density irregularities are correlated to backscatter observations from regions of auroral precipitation (Lund *et al.*, 1967), and (6) the very sharp low latitude hiss boundaries observed indicate ducting of the VLF energy owing to large electron density gradients perpendicular to the magnetic field lines.

3. Generation of Auroral Hiss

Concerning the region in space in which hiss is generated, the experimental work carried out so far seems to be consistent with the following picture: hiss is generated in the auroral regions of the magnetosphere, and the mechanism of generation is closely connected to very intense fluxes of electrons with energies of the order of a few keV. Since the energy flux of the energetic electrons is many orders of magnitude larger than the electromagnetic energy flux (Gurnett and O'Brien, 1964), it seems reasonable to assume that hiss is generated by the energetic electrons.

Mechanisms for conversion of the kinetic energy of the particles to electromagnetic energy can be divided into two main categories depending on whether the particles' longitudinal or transversal motion is the controlling factor. Mechanisms which depend on the longitudinal motion include Čerenkov radiation and a kind of amplification similar to the one going on in a traveling wave tube. The particles' transversal motion in the earth's magnetic field is almost circular, and the radiation connected to it is of the cyclotron type.

During the last decade several mechanisms have been proposed as explanations of VLF and LF emissions, but none of them has ever been proved. Reviews of VLF emission theories have been given by Helliwell (1965, 1966).

One of the more 'popular' mechanisms was the Čerenkov process, which also was the first mechanism ever suggested as an explanation of VLF emissions (Ellis, 1957), but Ellis, Liemohn (1965), and others concluded that the total energy produced by an incoherent Čerenkov radiation process was several orders of magnitude too low to explain the observed power densities.

In view of the better information we have today, compared with the situation a few years ago, on the magnetospheric environment, as well as on the occurrence in space and the intensity of VLF and LF emissions, it is clear that there exist several reasons for the discrepancies between theory and observations in earlier attempts to explain emissions by Čerenkov radiation. For example, no data on the thermal electron concentration above the F2-layer maximum were available when Ellis did his work in 1957, and only little was known about the population of energetic particles in the magnetosphere.

So the models used in earlier work considering Čerenkov radiation as a mechanism for VLF and LF emissions probably were unrealistic, and a new attempt was considered worthwhile by the author (Jørgensen, 1968), who discussed a model for a region in space in which the auroral hiss is believed to be generated. It was shown that the total power generated in this region is comparable to the observed power, and it was concluded that auroral hiss may be generated by incoherent Čerenkov radiation from electrons with energies of the order of 1 keV.

References

Carpenter, D. L.: 1966, *J. Geophys. Res.* **71**, 693.
Ellis, G. R.: 1957, *J. Atmosph. Terrest. Phys.* **10**, 303.
Feldstein, Y. I.: 1966, *Planetary Space Sci.* **14**, 121.
Flint, R. B.: 1968, 'Some Relationships between Aurora and Natural VLF and LF Noise at Plateau Station, Antarctica', Stanford Electronics Laboratories, Stanford University, Stanford, Calif.
Gallet, R. M.: 1959, *Proc. IRE* **47**, 211.
Gurnett, D. A.: 1966, *J. Geophys. Res.* **71**, 5599.
Gurnett, D. A. and O'Brien, B. J.: 1964, *J. Geophys. Res.* **69**, 65.
Helliwell, R. A.: 1965, *Whistlers and Related Ionospheric Phenomena*, Stanford University Press, Stanford, Calif.

Helliwell, R. A.: 1966, paper presented at URSI General Assembly, Munich, Germany, September 1966.

Jørgensen, T. S.: 1966a, *J. Geophys. Res.* **71**, 1367.

Jørgensen, T. S.: 1966b, 'Observations of the VLF Emission Hiss at 8 kc/s in Greenland 1964', Ionosphere Laboratory, The Technical University of Denmark, Lyngby, Denmark.

Jørgensen, T. S.: 1968, *J. Geophys. Res.* **73**, 1055.

Jørgensen, T. S. and Bell, T. F.: 1968, paper presented at NATO Advanced Study Institute, Plasma Waves in Space and in Laboratory, Røros, Norway, 17–26 April, 1968.

Jørgensen, T. S. and Ungstrup, E.: 1962, *Nature* **194**, 462.

Liemohn, H. B.: 1965, *Radio Sci.* **69D**, 741.

Lund, D. S., Hunsucker, R. D., Bates, H. F., and Murcray, W. B.: 1967, *J. Geophys. Res.* **72**, 1053.

Martin, L. H., Helliwell, R. A., and Marks, K. R.: 1960, *Nature* **187**, 751.

Morozumi, H. M.: 1962, 'A Study of the Aurora Australis in Connection with Association between VLF Hiss and Auroral Arcs and Bands observed at the South Geographical Pole 1960, SUI 62-14', State University of Iowa, Iowa City.

Starkov, G. V. and Fel'dshteyn, Ya. I.: 1967, *Geomagnetism and Aeronomy*, **7**, 48.

DIAGNOSTICS OF THE PARAMETERS OF THE
MAGNETOSPHERE AND OF THE INTERPLANETARY SPACE
BY MEANS OF MICROPULSATIONS

V. A. TROITSKAYA and A. V. GUL'ELMI

Institute of Physics of the Earth, Academy of Sciences of the U.S.S.R., Moscow, U.S.S.R.

Abstract. Systematic investigations of micropulsations, which were begun in connection with the IGY, showed a great variety of types. It was revealed that pulsations contain useful information about outer space. Observations over a wide network of stations allow conclusions to be drawn on the location and movement of the boundary of the magnetosphere, the concentration of cold plasma at great distances from the earth's surface, the energetic particles, the non-stationary processes developing in the geomagnetic trap during magnetic storms, etc. New possibilities of diagnostics of some of the parameters of the interplanetary space are suggested (e.g. mean velocity of the solar wind and structure of the interplanetary magnetic field). Methods of diagnostics and the results obtained are evaluated.

1. Introduction

In the series of successful achievements and discoveries of the last decade in cosmic space research, the results of investigations of geomagnetic pulsations possibly hold a modest but significant place. Systematic observations over a worldwide network of stations, as well as particular experiments, brought about the discovery of a great variety of micropulsations and led to the study of the basic regularities of their generation and distribution. Definite progress was made in understanding the physics of geomagnetic micropulsations and their relation to phenomena in the magnetosphere. Observations of geomagnetic micropulsations may well become one of the methods of diagnostics of cosmic space (Troitskaya and Gul'elmi, 1967). The great volume of information concentrated in micropulsations is due to the fact that their properties are connected with the structure of the magnetosphere, and with the phenomena occurring both inside and outside the magnetosphere. The analysis of geomagnetic micropulsations may usefully complement the direct satellite methods of investigating cosmic space.

Below, different aspects of diagnostics problems are discussed and preliminary results are given of determinations of different parameters. Some questions concerning the physical interpretation of several types of pulsations are also discussed.*

2. The Boundary of the Geomagnetic Cavity

The possibility of determining the location of the magnetospheric boundary arises

---

* In recent years several surveys on geomagnetic micropulsations have appeared (Troitskaya, 1964, 1967; Campbell, 1967). The general principles and the results of the diagnostics were described in surveys of Troitskaya and Gul'elmi (1967, 1968b).

from the following: the pulsations of Pc 2–4 types are identified as eigenmodes of hydromagnetic oscillations of the geomagnetic cavity. Their period consequently depends on the dimensions of the resonator. Owing to the very complicated structure of the resonance system, it is rather difficult to establish theoretically the connection of the oscillation spectrum with the location of the magnetospheric boundary. An empirical method of resolving the problem is therefore quite justified. With this aim in view, Pc 2–4 periods (T) observed on the sunlit side of the earth were compared with data on the location (R) of the magnetospheric boundary on the sun–earth line (Troitskaya and Bolshakova, 1964; Bolshakova, 1965).

The relation between T and R can be represented in the following form:

$$T = T_0 (R/10)^v \qquad\qquad (1)$$

where R is expressed in earth radii. This analysis of data (Explorer-12, Electron-4 and IMP-1) gives the following values of parameters: $T_0 \approx 28$ sec, $v \approx 4.5$.

The direct method of determining the location of the magnetospheric boundary is based on measurements by magnetometers and particle counters on board satellites crossing the surface of the geomagnetic cavity. It is observed, however, that between two successive crossings the location of the boundary can unpredictably change. Therefore, this method has rather low resolving possibilities in time.

While indirect methods of locating the magnetospheric boundary by means of pulsations are less precise, they make it possible to trace it continuously. Roederer et al. (1968) have elaborated an interesting method using data from a magnetometer installed on the geostationary satellite ATS-1. These measurements have much greater

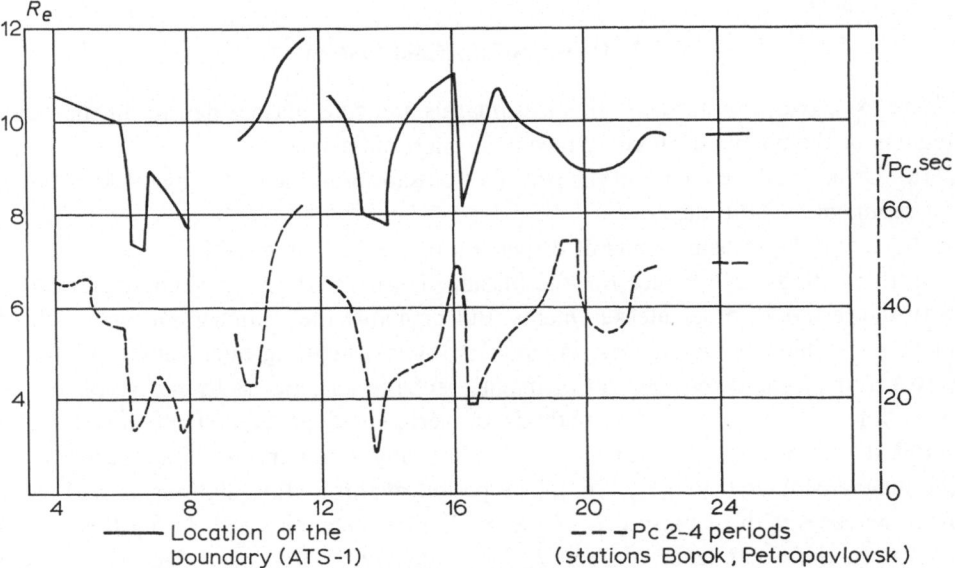

Fig. 1.   The relation of Pc 2–4 periods (lower curve) to the location of the magnetospheric boundary on the sun–earth line (upper curve).

resolving possibilities in time than the direct measurements on satellites with a strongly elongated orbit. Figure 1 shows a comparison of the boundary location (R) obtained on ATS-1 with the periods T of pulsations recorded at Borok and Petropavlovsk. The figure shows a good correlation of these quantities.

The possibility of tracing the location, and of following the movements of the sunlit magnetospheric boundary, has significance for investigations of the pulsations themselves. After positive or negative impulses of the magnetic field (S_i), the period of Pc 2–5 pulsations changes: i.e. it increases after S_i^- and diminishes after S_i^+ (Hirasawa *et al.*, 1966; Troitskaya *et al.*, 1967). It is interesting to note that after some S_i^-, Pc 2–4 disappear, that is, their amplitude falls below the noise level. Disappearance is the more probable, the greater is the absolute value of the impulse ΔB, and the greater the period T of pulsations before the S_i^-. Using the relation (1) and taking a suitable model of the sudden impulse it is possible, from the values of periods (T) and of ΔB, to estimate the location of the boundary after S_i^-. It has been found that pulsations disappear when after S_i^-, the 'radius' of the magnetosphere is greater than some critical value, equal approximately to 12.5 earth radii. In the paper of Troitskaya *et al.* (1968b) a description and a qualitative interpretation of this effect are given.

Some papers (Rostoker, 1967; Raspopov, 1968; Troitskaya and Schepetkov, 1967) analyse the connection between the open-closed line boundary on the night side and the spectrum of Pi 2. The statistical analysis shows that the Pi 2 period diminishes with increasing magnetic activity (and simultaneously the spectrum of oscillations broadens).

Further investigations tend to compare the location of the auroral oval and of the boundary of the outer radiation belt with the spectrum of Pi 2 pulsations (Troitskaya *et al.*, 1968f).

3. Density of the Cold Plasma

There exist two ground methods of estimating the cold plasma density in the outer regions of the magnetosphere by means of micropulsations.

(a) Using the dependence of the period of pulsations in the Pc 5 range (150–600 sec) on geomagnetic latitude.

(b) Using the measurements of dispersion in the Pc 1 range (0.2–5 sec).

Both methods give values for the plasma density that are in general agreement with the results of other measurements. But probably the main significance of these methods is that they open new possibilities for investigating micropulsations themselves. Indeed the determination of magnetospheric parameters by means of micropulsations is among the best methods of verifying their theoretical models. The simplest model for Pc 5 pulsations is the spherically symmetric toroidal oscillation of the magnetosphere (Dungey, 1963). The period of toroidal oscillations as well as of Pc 5 increases with increasing L of the oscillating magnetic shell. Using this model and given the dependence of T on L, Obayashi (1958) has constructed the vertical profile of plasma density in the plane of the geomagnetic equator. Other authors have later made similar calculations. Their results are presented in Figure 2 [(1) – Obayashi,

1958; (2) – Ol', 1963; (3) – Kitamura, 1965; (4) – Jacobs and Kitamura, 1966; (5) – Gul'elmi, 1965 (see the survey paper of Troitskaya and Gul'elmi, 1967)]. In spite of a rather great scatter, these results are in general agreement with the data obtained on satellites [(6) – Gringauz (Luna-2), 1961; (7) – Obayashi, OGO-A, 1965a; Slyš, Zonde-2, 1965].

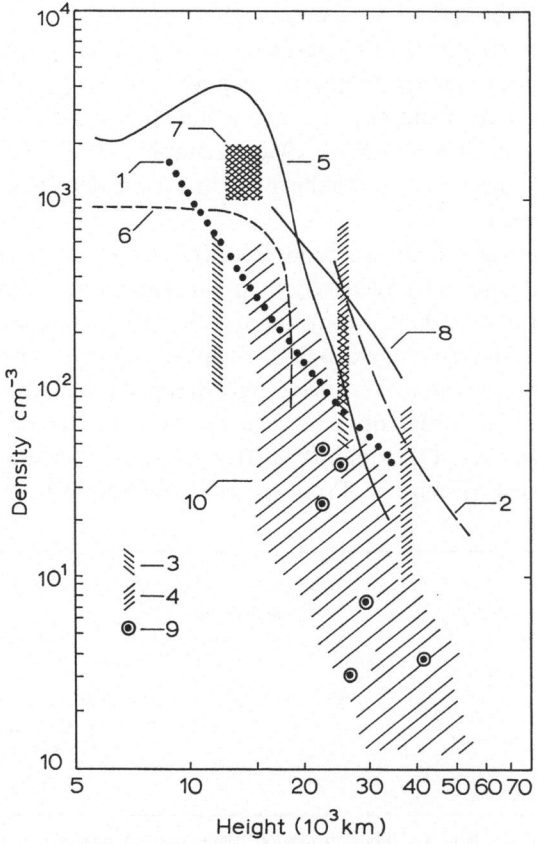

Fig. 2.   Distribution of cold plasma in the magnetosphere (see text).

The difficulty in using Pc 5 is the following. First, Pc 5 wavelengths are comparable with the dimensions of the magnetosphere. Therefore, in this case, computer calculations are required (see, e.g., Kitamura, 1965). Second, only the average dependence of T on L is known, and therefore the density profiles represent distributions averaged over a great time interval. There are, however, some special cases for which it is possible to obtain the 'instantaneous' dependence of $T(L)$. Brooks (1967), for instance, uses observations of radio auroral echo pulsations connected with sudden impulses.

The wavelengths of Pc 1 pulsations (pearls) are not very long. Therefore, with some assumptions the ray optic approximation may be used. According to the theory suggested by Jacobs and Watanabe (1965) and Obayashi (1965b) pearls are packets of ion-cyclotron waves (L mode) propagating along the geomagnetic field lines. Density determination is based on the fact that the dispersion of pearls (that is the dependence of the period of repetition on the frequency) is the greater, the closer the wave frequency to the proton gyrofrequency and the greater the plasma density at the top of the trajectory (Dowden and Emery, 1965; Watanabe, 1965; Wentworth, 1966; Liemohn *et al.*, 1967). The results of diagnostics are shown in Figure 2 [(9) – Wentworth, 1966; (10) – Liemohn *et al.*, 1967].

Let us now draw attention to two peculiarities of pearl propagation. First, the ratio of the wave frequency (carrier frequency of pearls) to the gyrofrequency at the top of the field line changes from case to case within the limits $\omega/\Omega_p \sim 0.3\text{–}0.7$, and its average value is ~ 0.5 (Wentworth, 1966; Liemohn *et al.*, 1967). Second, pearls propagate mainly in the region of the sharp fall in plasma density near the plasmapause (Liemohn *et al.*, 1967).

It seems that these two facts are connected. Indeed, the measurements carried out on satellite OGO-A showed a relatively high concentration of He^+ ions in the magnetosphere (Taylor *et al.*, 1965). According to calculations (Dowden, 1966; Gul'elmi, 1966) the L-wave with frequency $\omega > \Omega_{He^+}^{min}$ propagating from one conjugate point to the other in a proton-helium plasma is strongly damped (~ 54 db) if the concentration of helium ions is of the order of $\sim 1\%$. To reconcile this result with the results of dispersion measurements of the quantity ω/Ω_p we have to conclude that the trajectory of pearls is 'combined' (Figure 3). From the conjugate point the signal propagates in

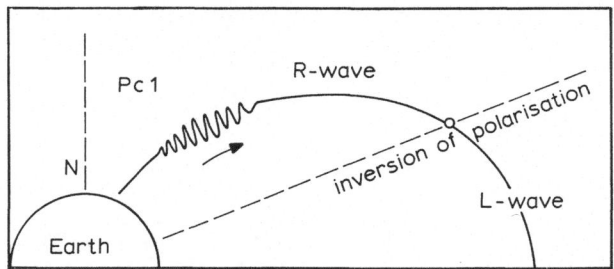

Fig. 3.   The 'combined' trajectory of pearls.

the form of a packet of L-waves. At the point along the trajectory where the carrier frequency of the signal $\omega \sim \Omega_{He^+}$ inversion of polarisation takes place, and the signal travels the rest of the path to the earth in the form of an R-wave.* The channeling of waves in this case is accomplished not directly by the geomagnetic field, but by plasma irregularities elongated around the field lines.

If the hypothesis of the 'combined' trajectories is true, then some corrections are required to derive the plasma density. The calculations show, however, that for $\omega/\Omega_p \geqslant 0.4$ the corresponding correction to the law of pearl dispersion is not very significant (Gul'elmi, 1968).

* The attention of the authors was drawn to this possibility by Prof. Obayashi (1966) after his reading of the paper by Gul'elmi (1966).

This interpretation clarifies somewhat why pearls are observed mainly in the region of the plasmapause: for $\omega/\Omega_p \sim 0.5$, the signal travels part of the way to the earth in the form of R-waves, the magnetic focusing of which is only possible owing to the existence of transverse gradients of plasma density. The reason why $\omega/\Omega_p \sim 0.5$ is closely connected with the mechanism of pearl generation, which will be discussed in the next section.

4. The Energetic Protons

Let us assume that pearls are generated near the equatorial plane by cyclotron instability of energetic protons (see survey papers by Troitskaya and Gul'elmi, 1967, 1968b). Using the results of dispersion measurements of the quantity ω/Ω_p it is possible to find the parameter L of the region where generation takes place. Now if one combines the data on ω/Ω_p and L with the condition of resonance of protons with L-waves,

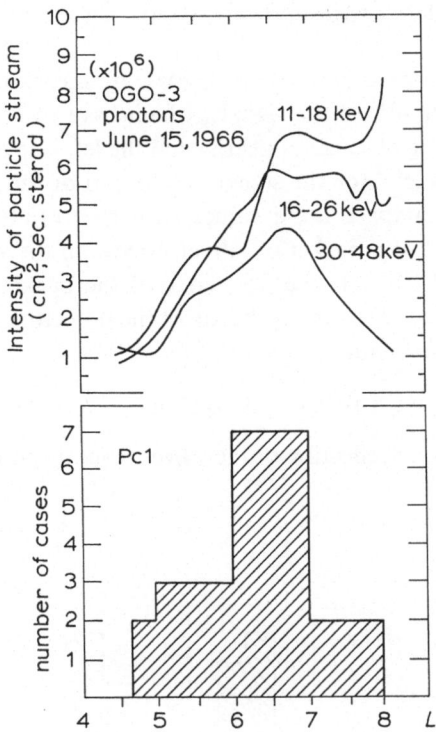

Fig. 4.   Pearl generation rate vs. L (lower part). The upper part of the figure shows the energetic proton intensity vs. L (Frank, 1967).

the energy E_p of the particles responsible for the generation of pearls can be estimated.

There exists also another, independent method of estimating the quantities ω/Ω_p, L and E_p from the data on the change in the spectrum of pearls that occurs simultaneously with sudden impulses S_i and sudden commencements SSC. The approximate expression for L, for instance, has the form:

$$L \approx 5.7 \left[\frac{\Delta\omega}{\omega} \frac{10^2}{\Delta B\gamma} \right]^{1/3}, \tag{2}$$

where $\Delta\omega$ is the change in the frequency of pearls, ΔB the value of the sudden impulse in gammas. Figure 4 shows the distribution of L shells found by this method (Troitskaya et al., 1968a). It is very important that the values of ω/Ω_p obtained in this way are also close to $\sim 1/2$. Therefore, the conclusion that the frequency of pearls is higher than the gyrofrequency of He$^+$ ions at the top of the trajectory is obtained by two independent methods, and consequently may be considered as firmly established. The scatter in energy of resonance protons is rather great:

$$E_p \sim 0.5 - 15 \text{ keV}.$$

To estimate the intensity of the stream of the resonance particles, the dynamics of the development of isolated series of pearls is investigated in the paper of Troitskaya et al. (1968c).

On the right of Figure 5 results are shown of measurements of the maximal amplification coefficient $A^{(1)}$ of the signals, which have twice traversed the region of generation (the beginning of series of pearls). On the left side is shown the distribution of damping coefficients $A^{(2)}$ for the signals at the end of the series.

It is interesting to compare these values with the results of calculations of the amplification coefficient of a packet of Alfvén waves on the way from one conjugate point to the other and back. If the frequency of the packet is significantly smaller than the gyrofrequency of He$^+$ ions at the top of the trajectory, for instance $\omega/\Omega_p \sim 0.1$, then in the linear approximation

$$A \text{ db} \approx 10^{-5}(L/10)^4 I\Delta - 40 \text{ bg}(1/p), \tag{3}$$

where p is the coefficient of reflection of the waves from the ionosphere, I is the density

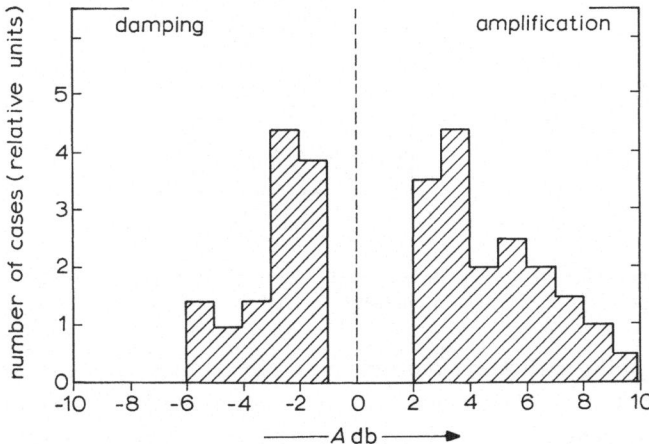

Fig. 5.   Distribution of the coefficients of amplification and of damping of pearls.

of the stream of resonance protons ($cm^{-2} sec^{-1}$). In the case of a two-temperature Maxwell distribution function of energetic protons, $\Delta = (T_\perp/T_\parallel - 1)$ is the degree of temperature anisotropy. In the case of an almost monoenergetic stream, $\Delta = (u/W)^{3/2}$ is the degree of 'monoenergeticity' (u is the average velocity of the stream and W is the scatter in velocities). Assuming that the measured values of $A^{(1)}$ must be of the order of (3) and $A^{(2)} \sim 40 \lg(1/p)$ and taking into account that $A^{(1)} - A^{(2)} \sim 10$ db we find that $I\Delta \sim 4 \times 10^7 cm^{-2} sec^{-1}$ and $I\Delta \sim 4 \times 10^6 cm^{-2} sec^{-1}$ for $L \sim 4$ and $L \sim 7$, respectively.

The condition $\omega/\Omega_p \sim 0.1$ is fulfilled in the region of the maximum of the outer radiation belt ($E \sim 100$ keV). In this region the density of the proton stream and the degree of anisotropy of their distribution on pitch angles are great enough for generation of Alfvén waves with the observed values of the amplification coefficient. Nevertheless, analysis shows that the pearls are generated in the spectrum range $\omega/\Omega \sim 0.5$.

This and other facts can be explained by the suggestion about the generation and evolution of series of pearls made by Troitskaya and Gul'elmi (1968c). Contrary to the generally accepted point of view, that pearls are generated as a result of instability of distribution of energetic particles with respect to small disturbances, it is suggested that the generation of pearls be considered as a principally nonlinear process. A mechanism of nonlinear instability is considered, which can take place only in a multicomponent nonhomogeneous plasma. According to this hypothesis, the presence in the magnetosphere of a small quantity of He^+ ions is a determining factor for pearl generation.

The series of oscillations are generated in the range $\Omega_{He^+} < \omega < \Omega_p$ and the linear stage is absent in the evolution of series. This is confirmed by observations (Troitskaya and Gul'elmi (1967)).

5. The Impulsive Processes in the Tail

The pulsations of the Pi 1–2 types having the character of noise are typical for the auroral zones. These irregular pulsations reveal a great variety of wave forms and usually occur simultaneously with X-ray pulsations in the stratosphere having similar spectra, with changes in the intensity of aurora, etc. (Barkus and Rosenberg, 1965; Campbell, 1967).

The whole complex of such phenomena is due to the fluctuating streams of electrons which are accelerated probably in the neutral sheet of the magnetospheric tail, and are injected in the ionosphere of auroral zones along the geomagnetic field lines. Therefore, the analysis of irregular magnetic pulsations provides indirect information on the processes in the geomagnetic tail.

We draw attention to the interesting type of irregular pulsations having the form of separate short bursts. These impulsive bursts have a rather wide spectrum, and their most characteristic feature is the tendency to occur in groups consisting of 3–5 bursts separated by time intervals of 10–20 min. Figure 6 shows the record and the sonogramme of bursts of irregular micropulsations, observed at conjugate points Sogra-Kerguelen (Gendrin and Troitskaya, 1967). The figure shows the high-frequency part of the bursts of micropulsations belonging to the type Pi 1. Often (but not always) the Pi 1 burst occurs simultaneously with longer-period oscillations of the Pi 2 type.

Comparison of the records of pulsations with the results of satellite measurements in interplanetary space showed that the first burst of irregular pulsations is often stimulated by the change in the N–S component of the interplanetary magnetic field, whereas the following bursts develop when the interplanetary field is relatively quiet

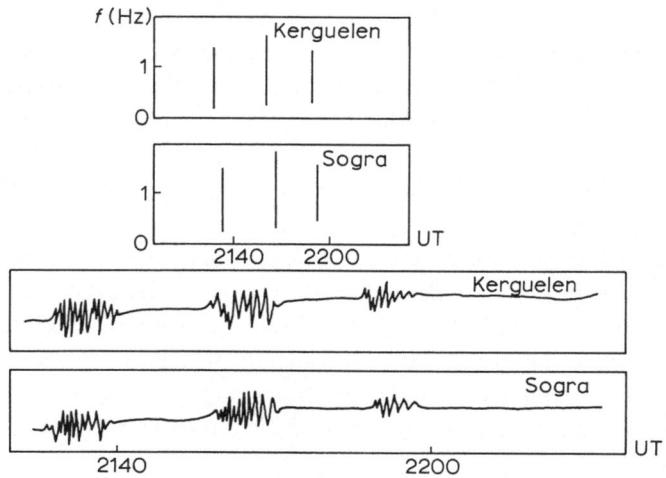

Fig. 6.   The impulsive bursts of irregular pulsations.

(Troitskaya *et al.*, 1968d). The period of repetition of the bursts is approximately equal to the time necessary for an isolated impulse or a weak shock wave to travel from the neutral sheet to the side boundary of the magnetospheric tail and back. As a result of disconnection of the field lines, two weak shock waves (or isolated impulses) may be generated, which propagate from the region of disconnection to the periphery of the tail. Being reflected from the side boundary of the tail, the waves begin to move toward each other. The collision of the shock waves in the vicinity of the neutral sheet stimulates a new disconnection of the field lines. Regeneration of shock waves leads to the multiple repetition of the effect. Therefore, the periodic bursts of magnetic pulsations are a 'far consequence' of shock-wave collisions of the magnetospheric tail.

It seems probable that the collision of the shock waves in the region lower or higher than the neutral sheet will be less effective than that directly in the sheet. Therefore, the multiple repetition of the bursts is more probable when two waves moving toward the axis of the tail reach the neutral sheet simultaneously. Since the North and South parts of the tail are more symmetrical at equinoxes, one may expect more frequent occurrence of multiple bursts of micropulsations in these periods. A preliminary analysis confirms this suggestion (Troitskaya *et al.*, 1968c).

Let us stress that we could not find the mechanism which would link the equinoctial maximum of the frequency of multiple burst occurrence to the helio-latitude of the earth.

6. Large Scale Electric Field

The direct measurements of the electric fields in the plasma of cosmic space encounter significant technical difficulties which will undoubtedly require much time to be resolved (Boyd, 1967). Our knowledge of the electric field in the magnetosphere is the result either of theoretical analysis of the problem or of indirect ground observations (see Obayashi and Nishida, 1968). For instance, the distribution of the large scale electric field in the magnetosphere is deduced from the data on ionospheric current systems, assuming that the geomagnetic field lines are equipotential. Recently Carpenter and Stone (1967) have estimated the electric field formed during the polar magnetic disturbances using data on VLF emissions generated simultaneously.

It is also interesting to consider the possibility of estimating the intensity of the electric field in the outer regions of the magnetosphere from data on pulsations with nonstationary spectra (Troitskaya et al., 1968e). The most typical pulsations of this kind are the intervals of oscillations in the Pc 1–Pi 1 range with slowly increasing frequency (IPDP, Figure 7). These pulsations are generated simultaneously at conjugate points during the main phase of a magnetic storm and are usually observed before midnight (Troitskaya, 1964, 1967; Gendrin et al., 1967; Heacock, 1967; Kenney and Knaflich, 1967).

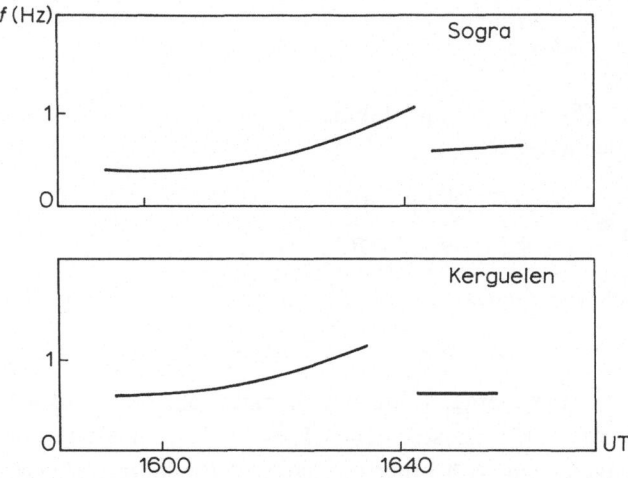

Fig. 7.   The interval of pulsations of diminishing periods (IPDP).

The frequency increase of pulsations occurs simultaneously with intensified activity of the auroral electrojet, the appearance of aurora in middle and low latitudes, variations of the structure of the belt of energetic electrons ($\geqslant 100$ keV), etc.

It seems quite probable that the occurrence of IPDP is intrinsically connected with the generation in the magnetosphere of the large-scale slowly varying electric fields responsible for the polar substorm.

The mechanism of IPDP generation is still unclear. We shall assume that these pulsations are generated near the equatorial plane as a result of cyclotron instability of energetic proton distributions. Thus at each moment of time the average frequency of oscillations is in some definite ratio $\eta < 1$ to the gyrofrequency of protons at the top of the field line

$$\omega = \eta \Omega_p. \tag{4}$$

If the parameters of the medium in the region of generation (on which η and Ω_p depend) change slowly in time, then the spectrum of the generated frequencies (4) will change. Let us assume that the main reason for the frequency increase in the interval IPDP is the radial drift of the oscillators inside the magnetosphere which is due to the action of the azimuthal electrical field E_φ:

$$\dot{\omega} \approx \frac{l}{m_p a} \frac{E_\varphi}{\Omega_p} \frac{\partial}{\partial L} (\eta \Omega_p), \tag{5}$$

where e and m_p are the proton charge and mass, L the parameter of the magnetic shell, and a the radius of the earth. In order to solve this problem we shall make two simplifying assumptions:

$$\left| \frac{\partial \ln \eta}{\partial \ln \Omega_p} \right| \ll 1 \tag{6}$$

and

$$\Omega_p = \Omega^{(0)} L^{-\mu}. \tag{7}$$

Combining (4), (5), and (7) and taking into account (6), we obtain the following formula for calculation of E_φ:

$$E_\varphi \approx - \frac{m_p a}{e \mu \eta} \left[\eta \frac{\Omega_p^{(0)}}{\omega} \right]^{1/\mu} \frac{d\omega}{dt}. \tag{8}$$

For preliminary estimates we shall take the following values of the parameters:

$$\Omega_p^{(0)} \approx 3 \times 10^3 \text{ rad/sec}; \quad \mu \approx 3 \text{ (dipole field)}.$$

For a typical case the average frequency of pulsations $\omega \approx 6$ rad/sec and the drift of the frequency $\approx 5 \times 10^{-3}$ rad/sec^2. Then $E_\varphi \sim 10^{-5}$ V/cm. This value is in good order of magnitude agreement with other estimates of the large scale electric field on the periphery of the closed cavity of the night magnetosphere (see, for instance, Block, 1967; Boyd, 1967; Carpenter and Stone, 1967). It is clear, of course, that each of these assumptions must be carefully checked. Assumptions (4) and (6) seem justified, but the precise value of the parameter η must be found in a full theory of IPDP generation. The validity of (7) seems rather doubtful. To establish the precise dependence $\Omega_p(L)$ it is necessary to use the model of the geomagnetic field which takes into account the ring current RC. It is quite probable that owing to variations of RC in time it will be necessary to take into account terms of the $\partial\Omega_p/\partial t$ and $\partial\eta/\partial t$ type, which were not considered in (5).

We are fully aware that this method of estimating the electric field in the magnetosphere can only complement the direct measurements, and that in the very rough form described it cannot stand comparison with other ground methods. Nevertheless, after precise verification of the above assumptions, the analysis of IPDP spectra can be useful.

7. Parameters of Interplanetary Space

Almost the whole complex of physical processes in the magnetosphere is due to changes occurring in the interplanetary space surrounding the earth. One can say that the magnetosphere is a giant cosmic probe which reacts very sensitively to changes in the parameters of interplanetary space. From the point of view of a specialist in the diagnostics of the interplanetary plasma by means of geomagnetic pulsations, the investigation of the magnetosphere is a method of analysis of the properties of this giant 'apparatus'.

There is much work to be done in this direction but the first successful investigations already allow us to formulate the main problems and to outline some preliminary results. First of all, the data on pulsations Pc 2–4 giving the location and the movements of the magnetospheric boundary provide indirect information on the dynamic pressure of the solar wind at the orbit of the earth. Because the density of the interplanetary plasma and its radial velocity V are probably connected in some definite way, there is a possibility of estimating the value of V by using the data on periods of Pc 2–4 (Bolshakova, 1965b). Direct comparison shows that the coefficient of correlation Pc 2–4 with the average velocity of the solar wind is significant ($\tau = 0.73 \pm 0.17$). The relation between these quantities has the form

$$U(\text{km/sec}) = (850 \pm 15) - (9.5 \pm 0.8)\, T(\text{sec}). \tag{9}$$

This formula can be used to estimate U during times when for one reason or another direct measurements are lacking.

An interesting possibility of gaining information about the general structure of the interplanetary magnetic fields from an analysis of Pc 2–4 and Pi 2 amplitudes was revealed recently (Bolshakova and Troitskaya, 1968; Schepetnov, 1968). The data on changes in the orientation of the interplanetary field obtained by IMP-1 were compared with the amplitude modulation of Pc 2–4 pulsations. It was found that there exists a definite orientation of the interplanetary field which is associated with the disappearance of these pulsations. Figure 8 shows the record of pulsations registered at Borok (upper part of the figure) and the changes in time of the projection of the interplanetary field on the ecliptic plane (lower part of the record, Bolshakova and Troitskaya, 1968). The pulsations disappear when the azimuthal projection of the interplanetary field is rising (B_r is the component of the field in the direction sun–earth, B_φ is the azimuthal component).

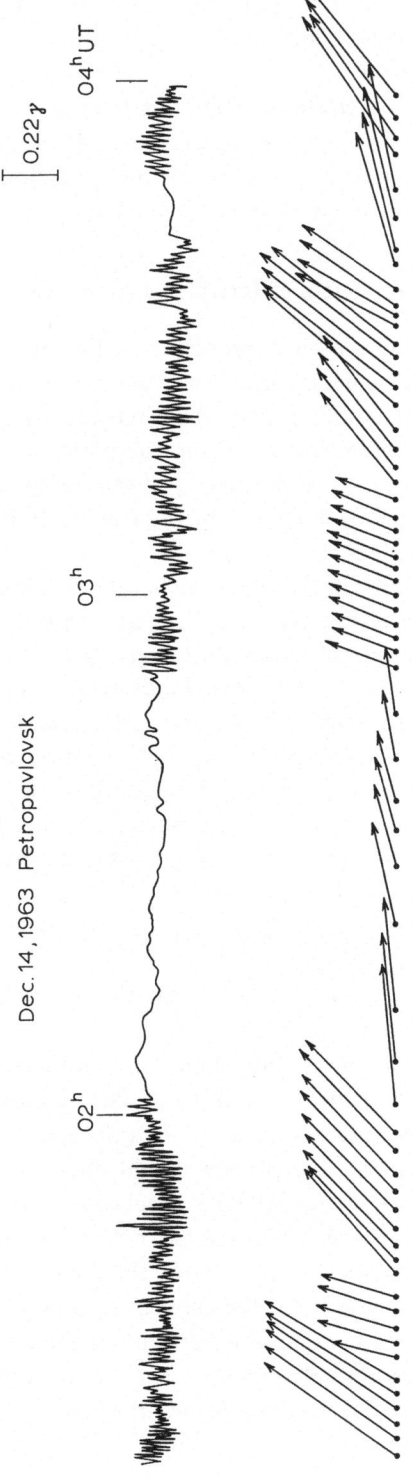

Fig. 8. Modulation of Pc 2–4 amplitudes by the interplanetary magnetic field.

The existence of such an effect can be understood qualitatively. Indeed, the Pc 2–4 pulsations are generated on the sunward surface of the magnetosphere. One of the possible mechanisms of their generation is an instability of the Kelvin-Helmholtz type. This instability arises from the flow of the solar wind around the magnetosphere (Dungey, 1963; Moskwin and Frank-Kamenetskii, 1967). But it is known that an interplanetary field of definite intensity and definite orientation can suppress the instability. A quantitative interpretation of the observations is difficult, largely owing to lack of data on the topology of the outer field near the magnetospheric boundary. Independently of the correctness of such an interpretation, the experimentally established correlation makes it possible to study irregularities of the interplanetary field by means of ground observations of geomagnetic pulsations. The structure of the interplanetary magnetic fields is rather complicated and includes irregularities of different dimensions and types.

The smoothed projection of the field lines on the ecliptic plane has the form of spirals, which at the orbit of the earth form an angle of $\psi \sim 50°$ with the radius vector from the sun.

As the sharp decrease in pulsation amplitude occurs in those cases in which the direction of the field becomes almost azimuthal, the angular scale of the irregularities is $\langle \Delta\psi \rangle \sim 40°$, i.e. these are large irregularities. One can estimate their dimensions by means of the graph in Figure 9. The full line (left scale) gives the distribution of

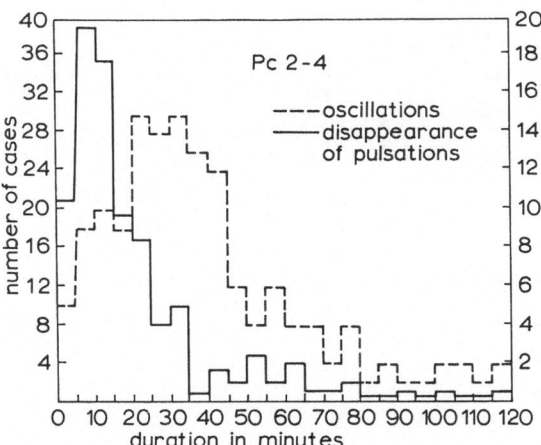

Fig. 9. The distribution of the intervals with records of Pc 2–4 pulsations (right scale) and intervals when pulsations are absent (left scale).

duration of the intervals of records when the pulsations are absent. The dotted line (right scale) gives the distribution of the intervals of records when the pulsations are present.

The average duration of intervals without pulsations is around 600 sec. The duration of intervals with pulsations is approximately 3 times greater. If we assume that the irregularities of the magnetic field pass the earth with the velocity of the solar wind, then the corresponding dimensions will be $l \sim 3 \times 10^{10}$ cm and $d \sim 9 \times 10^{10}$ cm, and the ratio of these quantities will be ~ 3. Irregularities estimated in this way appear to be just the same as those which determine the character of the angular distribution of solar cosmic rays at the first (anisotropic) stage of their appearance.

The night pulsations Pi 2 are sensitive to the component of the interplanetary field $B_\perp$, which is perpendicular to the ecliptic plane. As a rule, they appear more frequently after $B_\perp$ changes direction from North to South. The amplitude of Pi 2 is in some way connected with the value of $\Delta B_\perp$ (see Figure 10, Schepetnov, 1968). Further simultaneous ground and satellite observations will make this connection more definite.

8. Problems for Further Investigation

Almost every geophysical experiment is by its nature of global character. The investigation of geomagnetic micropulsations is one more example of this kind. The development of methods of determining the dynamic parameters of the magnetosphere requires special observations over a large part of the earth's surface, which are impossible without a wide international cooperation. Of special significance are observations of

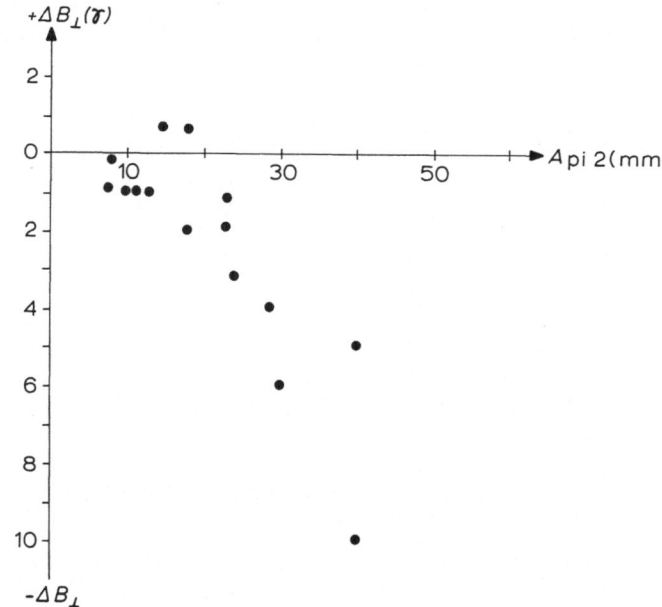

Fig. 10. The dependence of Pi 2 amplitude on the change in the North-South component of the interplanetary magnetic field.

pulsations at conjugate points near the geomagnetic poles and the equator. To clarify the distribution of pulsations on the earth's surface, observations on special networks of stations are important.

It is evident that investigation of micropulsations directly depends on coordination of ground observations with the observations by rockets and satellites.

So far, only the general features of the theory of pulsations have been outlined. A great deal of work has still to be done in elaborating the mechanism of their generation in order to work out a quantitative theory. Very important are:

(a) Further calculations of the spectrum of eigenmodes for realistic models of the magnetosphere.

(b) Calculations of the critical frequency, phase velocity and damping of the hydromagnetic waves in the ionospheric waveguide.

(c) Elaboration of the nonlinear theory of generation of geomagnetic micropulsations.

One of the present problems of method is the introduction of new indexes of activity based on geomagnetic micropulsations. The system of such indexes, together with the classical indexes of magnetic activity, will serve as a new tool for investigating solar-terrestrial relationships.

The variety of their types, their different properties, and the enigma of their origin make micropulsations one of the most interesting natural phenomena. This is why they first attracted the attention of geophysicists. In the course of investigations, somewhat unexpectedly, the high informative value of micropulsations was discovered, and the direction of research markedly changed. Future investigations of geomagnetic micropulsations will probably concentrate on interplanetary space parameters and large scale cosmic processes will be stressed.

References

Barkus, J. R. and Rosenberg, T. J.: 1965, *J. Geophys. Res.* **70**, 1707.

Block, L. P.: 1967, *Space Sci. Rev.* **7**, 198.

Bolshakova, O. V.: 1965a, *Geomagnetism i Aeronomija*, **5**, 865.

Bolshakova, O. V.: 1965b, *Astron. J. (USSR)*, **42**, 859.

Bolshakova, O. V. and Troitskaya, V. A.: 1968, *Dokl. Akad. Nauk SSSR* **180**, 343.

Boyd, R. L. F.: 1967, *Space Sci. Rev.* **7**, 230.

Brooks, D.: 1967, *J. Atmos. Terr. Phys.* **29**, 589.

Campbell, W. H.: 1967, in *Physics of Geomagnetic Phenomena*, Vol. II. (ed. by S. Matsushita and W. H. Campbell), Academic Press, New York, p. 821.

Carpenter, D. L. and Stone, K.: 1967, *Planetary Space Sci.* **15**, 395.

Dowden, R. L.: 1966, *Planetary Space Sci.* **14**, 1273.

Dowden, R. L. and Emery, M. W.: 1965, *Planetary Space Sci.* **13**, 773.

Dungey, J. W.: 1963, in *Geophysics: the Earth Environment* (ed. by C. De Witt, J. Hieblot, and A. Lebeau), Gordon and Breach, New York and London, p. 505.

Frank, L. A.: 1967, *J. Geophys. Res.* **72**, 1905.

Gendrin, R. and Troitskaya, V.: 1967, *Compt. Rend. Acad. Sci. France* 1.

Gendrin, R., Lacourly, S., Troitskaya, V., Gokhberg, M., and Shchepetnov, R.: 1967, *Planetary Space Sci.* **15**, 1239.

Gringauz, K. I.: 1961, 'Iskustvennie Sputniki Zemli', No. 12, 105.

Gul'elmi, A. V.: 1966, 'About Nature of hm Whistlers', Presented at Inter-Union Symposium of Solar-Terrestrial Physics, Belgrade, Yugoslavia.

Gul'elmi, A. V.: 1968, 'About Theory of the Diagnostics of Plasma Density in the Magnetosphere', *Geomagnetism i Aeronomija* (in press).

Heacock, R. R.: 1967, *J. Geophys. Res.* **72**, 399.

Hirasawa, T., Nishida, A., and Nagata, T.: 1966, *Rep. Ionosph. Space Res. Japan* **20**, 51.

Jacobs, J. A. and Kitamura, T.: 1966, *Trans. Amer. Geophys. Union* **47**, 73.
Jacobs, J. A. and Watanabe, T.: 1964, *J. Atmos. Terr. Phys.* **26**, 825.
Kenney, J. F. and Knaflich, H. B.: 1967, 'IPDP Events and Their Generation in the Magnetosphere', Boeing Sci. Res. Lab. D1-82-0616.
Kitamura, T.: 1965, *Rep. Ionosph. Space Res. Japan* **19**, 21.
Liemohn, H. B., Kenney, J. P., and Knaflich, H. B.: 1967, *Earth Planetary Sci. Letters* **2**, 360.
Moskvin, In. L. and Frank-Kamenetskii, D. A.: 1967, *Dokl. Akad. Nauk SSSR* **174**, 1079.
Obayashi, T.: 1958, *Ann. Geophys.* **14**, 464.
Obayashi, T.: 1965a, *Rep. Ionosph. Space Res. Japan* **19**, 2.
Obayashi, T.: 1965b, *J. Geophys. Res.* **70**, 1069.
Obayashi, T.: 1966, Private communication.
Obayashi, T. and Nishida, A.: 1968, *Space Sci. Rev.* **8**, 3.
Ol', A. I.: 1963, *Geomagnetism i Aeronomija* **3**, 113.
Raspopov, O. M.: 1968, *Geomagnetism i Aeronomija* **8**, 328.
Roederer, J. G., Cummings, W. D., Coleman, P. J., and Robbins, M. F.: 1968, 'Determination of Magnetospheric Parameters from Magnetic Field Measurements at Synchronous Altitudes', Univ. of Denver, 80210.
Rostoker, G.: 1967, *J. Geophys. Res.* **72**, 2032.
Scheptnov, R. V.: 1968, Dissertation, Moscow.
Slyš, V. I.: 1965, *Kosmich. Issled. (USSR)* **3**, 760.
Tayler, H. A., Brinton, H. C., and Smith, C. R.: 1965, *J. Geophys. Res.* **70**, 5769.
Troitskaya, V. A.: 1964, in *Research in Geophysics* (ed. by H. Odishaw), Vol. 1, M.I.T. Press, Cambridge, Mass., p. 485.
Troitskaya, V. A.: 1967, in *Solar-Terrestrial Physics* (ed. by J. W. King and W. S. Newman), Ch. VII, Academic Press, New York, p. 213.
Troitskaya, V. A. and Bolshakova, O. V.: 1964, 'Continuous Pulsations of the Earth's Electromagnetic Field', Presented at the Symposium of ULF Electromagnetic Fields, Boulder, Colo.
Troitskaya, V. A. and Gul'elmi, A. V.: 1967, *Space Sci. Rev.* **7**, 689.
Troitskaya, V. A. and Gul'elmi, A. V.: 1968a, 'Geomagnetic Pulsations and Diagnostics of the Magnetosphere', *Usp. Phys. Nauk (USSR)* (in press).
Troitskaya, V. A. and Gul'elmi, A. V.: 1968b, 'A New Hypothesis about Pearls Exitations' (in press).
Troitskaya, V. A. and Schepetnov, R. V.: 1967, 'Classifications of Pi 2-pulsations', presented at Inter-Union Symposium, St.-Galle, Switzerland.
Troitskaya, V. A., Schepetnov, R. V., and Gul'elmi, A. V.: 1968a, *Dokl. Akad. Nauk (SSSR)* 1.
Troitskaya, V. A., Schepetnov, R. V., and Gul'elmi, A. V.: 1968b, 'An Effect of Disappearance of the Pc 2–4-Pulsations', *Geomagnetism i Aeronomija* (in press).
Troitskaya, V. A., Matveeva, E. T., and Gul'elmi, A. V.: 1968c, 'Amplification and Attenuation of Pearls', *Geomagnetism i Aeronomija* (in press).
Troitskaya, V. A., Gul'elmi, A. V., and Schepetnov, R. V.: 1968d, 'Polar Substorms and Geomagnetic Pulsations', Presented at Inter-Union Symposium, Gagry, Black Sea.
Troitskaya, V. A., Schepetnov, R. V., and Gul'elmi, A. V.: 1968e, *Geomagnetism i Aeronomija* **8**, 798.
Troitskaya, V. A., Matveeva, E. T., Ivanov, K. G., and Gul'elmi, A. V.: 1968f, 'Change of Spectrum of the Pearls, *Geomagnetism i Aeronomija* (in press).
Troitskaya, V. A., Feldshtein, Ja. I., Schepetnov, R. V., and Raspopev, O. M.: 1968h, 'Variability of the Concentration of Charged Particles in the Magnetosphere', *Dokl. Akad. Nauk SSSR* (in press).
Watanabe, T.: 1965, *J. Geophys. Res.* **70**, 5839.
Wentworth, R. C.: 1966, *J. Geomagn. Geoelectr.* **18**, 257.

EFFECTS OF IONOSPHERIC MOTIONS AND IRREGULARITIES ON HF RADIO PROPAGATION*

T. M. GEORGES

ESSA Research Laboratories, Boulder, Colo., U.S.A.

Abstract. A tutorial account is given of observations of motions and irregularities in the ionosphere, of current theories of their origin, particularly those attributable to atmospheric-wave mechanisms, and of ray-tracing techniques for interpreting radio observations.

1. Introduction

This lecture is about the effects on high-frequency radio propagation of motions and irregularities in the structure of the ionosphere. The commonly used word 'irregularities' tends to convey the erroneous impression that there exists some 'normal' state of the ionosphere that is in some sense regular, stationary, or undisturbed, upon which are occasionally superimposed moving perturbations of various kinds. One often gets the initial impression, from displays of synoptic electron density profiles and such, that the ionosphere is essentially a static reflector of radio waves, changing somewhat from day to night, seasonally and with geographic location, but whose motions and small-scale structure are of negligible importance. The main impression I would like to leave is rather that the ionosphere is by nature a *dynamic*, or continuously time-varying, medium; that the various synoptic pictures of the ionosphere can be thought of as representing its actual structure in somewhat the same sense that the surface of the ocean can be called horizontal, and that one should not expect them to apply in anything more than a statistical or long-term-average sense. The ionosphere goes through periods of relative calm as well as through stormy and 'turbulent' phases, but never actually settles down to a perfectly calm or static state.

Anyone who has ever listened to a short-wave radio broadcast has experienced the effects of the dynamic ionosphere in the form of the apparently random fading that characterizes these signals; time variations in received signal strength must reflect time variations in the propagation medium. The communicator cannot normally adjust his operating parameters to 'follow' these variations, as he may be able to for longer-term changes, and has, in effect, to deal with a time-varying medium. Clearly, then, an understanding of the ionosphere as a telecommunications medium would be incomplete without a description of its dynamic behavior. Besides communications, the ionosphere is also relied upon as an indicator of 'space weather' and in various kinds of surveillance activities. Here it is clearly of value to know to what extent ionospheric sensors can give false indications because of natural ionospheric motions.

It turns out that, like the ocean surface, the dynamic behavior of the ionosphere is

* This lecture was originally prepared for the ESSA/University of Colorado Electromagnetic Propagation Course, June, 1968.

not so irregular that it has to be treated from a statistical point of view; indeed, there does seem to be some regularity in many kinds of so-called irregularities. If one were to Fourier-analyze the time behavior of ionospheric variations, say electron-density fluctuations at some fixed location, quasi-periodicities would be found, ranging from the 11-year sunspot cycle down to time scales of seconds and less. Figure 1 shows schematically how such a spectrum analysis might look. The ordinate N_{max}/N_{min} has been used to give a semi-quantitative estimate of the magnitude of electron density fluctuations to be expected.

The global-scale and long-term (longer than several hours) structural changes are now understandable and predictable to a certain extent. Good tutorial summaries of our state of knowledge in this area are given by Rishbeth (1967) and Rishbeth and Garriott (1964).*

Superimposed on the global-scale and long-term variations, however, relatively small-scale irregularities seem to exist at all levels in the ionosphere. There is growing evidence that the periodicities of many, if not most, of these motions can be explained in terms of waves in the neutral atmosphere interacting with the ionized constituents of the upper atmosphere. Our understanding of these processes is not yet good enough to permit predictions of the quality possible for long-term motions, but research in this area holds the promise of our being able to cope with short-term variations as well.

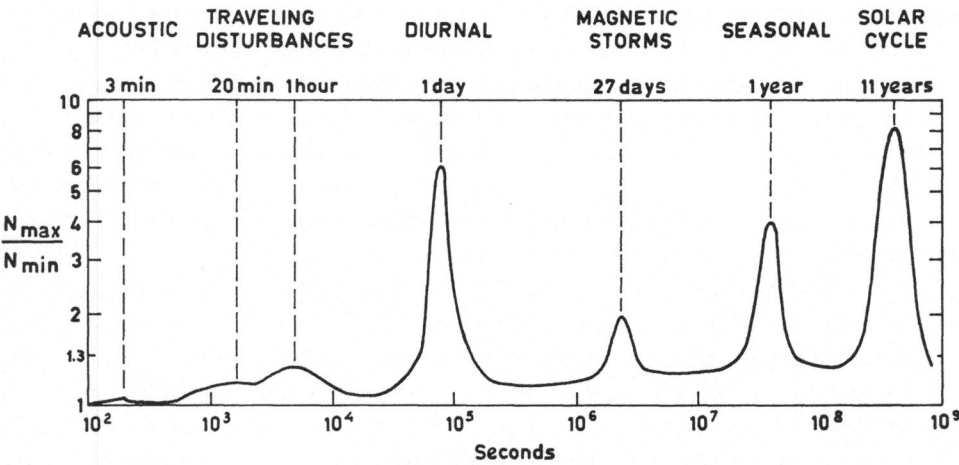

Fig. 1.   Schematic sketch of the spectral content of ionospheric motions.

The following sections outline: (a) dynamic characteristics of the ionosphere as seen by the major radio-sounding techniques; (b) some of the physical processes believed to be involved in the ionospheric effects of atmospheric waves; and (c) application of radio ray tracing and ionosphere modeling to the computer simulation of dynamic effects.

* The references cited here have been selected to provide good starting points for those wishing to pursue various points that can be only briefly touched upon in this summary.

2. Radio Observations of Ionospheric Motions

The way radio waves interact with the dynamic ionosphere can be partially understood by studying the results of the major radio-sounding techniques that have been used for that purpose. These techniques are:

(a) fading correlation (including satellite and radio-star scintillation);
(b) direction-of-arrival measurements;
(c) ionosondes (including spaced);
(d) fixed-frequency pulse sounding (including spaced);
(e) Thomson ('incoherent') backscatter;
(f) HF ground backscatter;
(g) Doppler and phase-path sounding (oblique, vertical and spaced).

The earliest experimental studies of short-term motions were measurements of signal-strength fading of downcoming waves at closely spaced (a few wavelengths) receivers. This technique, now called fading correlation, or 'ionospheric drift' measurement, determines the statistical properties of the radio-diffraction or 'shadow' pattern formed on the ground by a radio wave passing through an irregular ionosphere. An analogy is often drawn with the light pattern that sunlight forms on the bottom of a disturbed pool. Inferences are often made, now using sophisticated correlation-analysis schemes, about the structure and motions in the ionosphere that cause the diffraction patterns, but most of the inferences are still subject to considerable debate because of the multitude of processes that can cause radio fading. Apart from their use as an ionospheric diagnostic technique, fading-correlation measurements provide important information on the temporal and spatial correlation characteristics of ionospheric transmissions. An extensive review of this technique and its results is given by Briggs and Spencer (1954), and the results of a recent symposium on the significance of these measurements appears in the *Journal of Atmospheric and Terrestrial Physics*, May, 1968.

Several radio-astronomical and satellite variations of the fading-correlation scheme have recently appeared. Reception of signals transmitted from satellites, or coming from cosmic radio sources, at spaced receivers provides evidence of small (a few kilometers) moving irregularities, probably field-aligned, in the vicinity of the height of *F*-layer peak electron density. Faraday rotation and Doppler shift measurements on these signals show short-period (a few minutes) variations in the total electron content along the radio paths. Aarons (1963) provides a recent collection of research papers in the area of radio astronomical and satellite studies of the atmosphere, as does a recent special issue of *Radio Science* (October 1966). Booker (1958) gives a good introduction to the theory of scintillation effects. Although such measurements may demonstrate effects to be expected on earth-satellite communications circuits, they are of limited usefulness in probing the detailed structure of the ionosphere, because they measure effects that are integrated over long radio paths.

Another early technique was the measurement of fluctuations in the azimuth and elevation angles of arrival of ionospherically propagated signals. Apparently random fluctuations of fractions of a degree are common, and quasi-periodic fluctuations of a few degrees are occasionally observed. The most common periods of the large fluctuations seem to be in the tens of minutes. Early observations of these effects and deductions about the ionospheric motions that caused them were reported by Bramley (1953). The technique is seldom used today as a diagnostic tool, but the effect is of concern in radiolocation activities.

Sweep-frequency pulse sounders (ionosondes) have long been the principal tool for ionosphere research. When soundings are made at spaced locations or in rapid time sequence (every few minutes), information is often found about the motions and small-scale structure of the ionosphere. A time sequence of ionograms can be put together into a motion picture in which motions and irregularities can be clearly identified. A striking feature that is often seen in such movies is the appearance of 'kinks' or irregularities in the profile that travel down the F-layer trace from higher to lower virtual heights. Figure 2 illustrates a few steps in this process. These motions

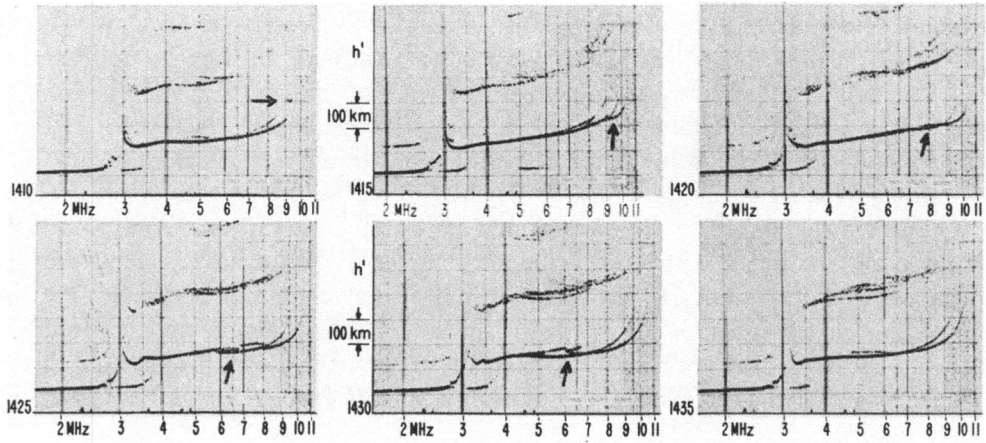

Fig. 2.   A time sequence of ionograms showing a traveling disturbance (arrows).

have been identified with irregularities that appear to be wavelike in character and have come to be called Traveling Ionospheric Disturbances (TIDs). Those commonly observed range in period from about ten minutes to over an hour, and involve electron density perturbations of 10 to 20%. The larger disturbances often accompany large geomagnetic storms, and these do not appear as kinks, but as vertical displacements of the ionogram almost as a whole. Plots of plasma-frequency contours in the virtual-height-vs.-time plane reveal the nature of these disturbances (Figure 3).

One of the great mysteries of ionospheric observations is a phenomenon called 'spread-F'. So-called because of the spread or smeared appearance of the F-layer echo on ionograms where it was first observed (Figure 4), the term is now commonly applied to the ionospheric phenomenon itself. Spread-F has since been associated to some extent with other similar ionospheric phenomena such as radio-star scintillations and spreading of the echo frequency in Doppler sounding. Some attempts have been made to associate spread-F with the rapid fading that sometimes appears on HF radio signals, but the results have been inconclusive. Though the detailed mechanisms are still unknown, these phenomena are believed to arise from scattering by, or at least a large number of coherent reflections from, small, field-aligned irregularities high in the F region. The irregularities may arise either from some kind of electrodynamic instability in the ionosphere or from short atmospheric waves. Much work remains before spread-F will be completely understood. A good survey of progress to date is given by Herman (1966).

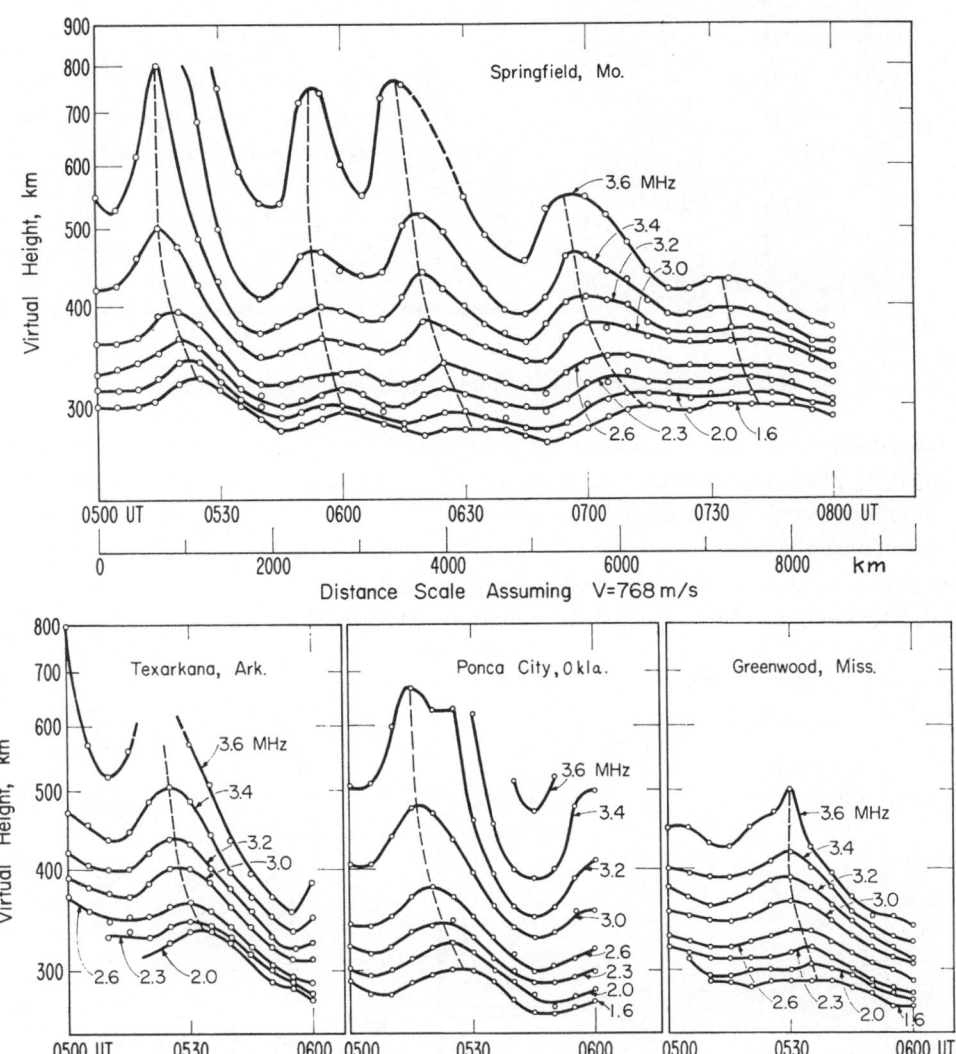

Fig. 3. A larger traveling disturbance found by constructing isoionic contours in the virtual-height-vs.-time (h', t) plane from sequences of ionograms taken 5 min apart and at geographically spaced locations.

Another common 'irregular' feature in the ionosphere is called 'sporadic-E'. A great deal of literature exists on the subject (Smith and Matsushita, 1962; and a special issue of *Radio Science*, February, 1966). Briefly, sporadic-E denotes the presence of thin layers of greatly enhanced ionization in the E region. Several kinds of sporadic-E layers have been classified, including equatorial, high-latitude, and mid-latitude. Mid-latitude sporadic-E ionization is probably caused by wind shears in atmospheric tides and gravity waves and probably involves slowly recombining metallic ions from meteor debris. Although the presence of the various species of sporadic-E layers can

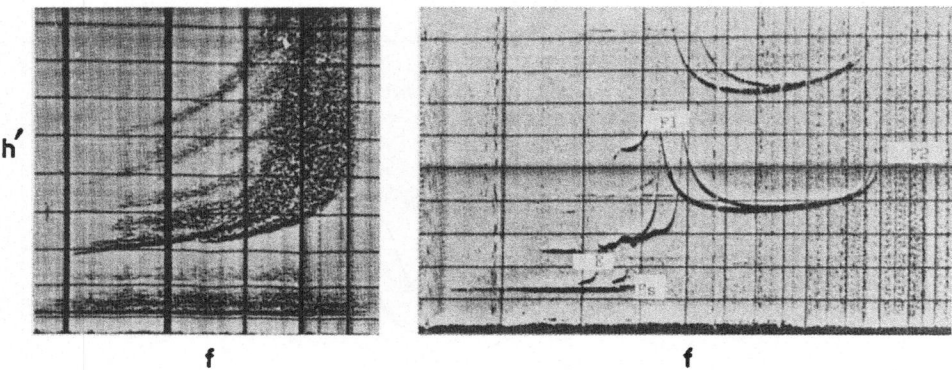

Fig. 4.   Ionograms showing spread-F (left) and sporadic-E (right).

have profound effects on communications circuits, through the introduction of entirely different propagation modes, no effective way of predicting its occurrence other than on an empirical basis has yet been devised.

Some of the labor of making (h', t) plots from a sequence of ionograms can be removed at the expense of variable-frequency information by recording h' vs. time directly using fixed-frequency pulse sounding. Munro (1958) used this technique with

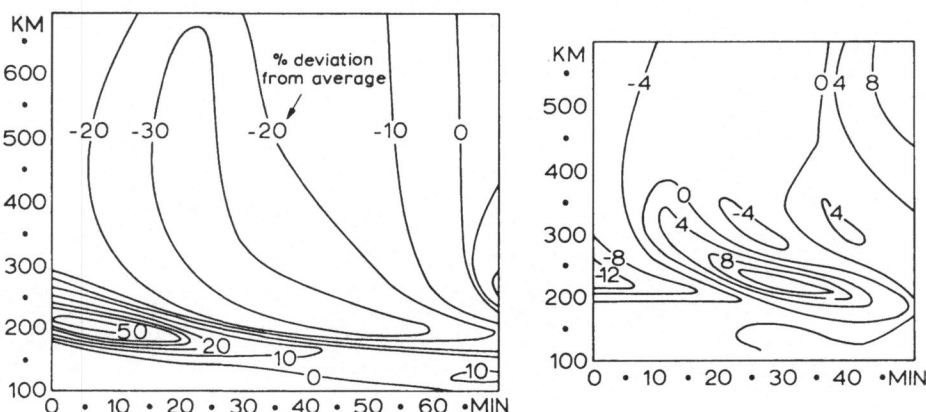

Fig. 5.   Height–time profiles of large traveling disturbances seen with an incoherent-scatter radar (after Thome).

spaced transmitters to compile impressive statistics on the occurrence of traveling disturbances with periods between about 10 and 30 min.

Thomson (or 'incoherent') scatter sounding has provided data on the structure of large irregularities, not only below but also above the F-layer peak of electron density which is inaccessible to ordinary sounders (Thome, 1964). Figure 5 shows electron density contours, in the height-time plane, for a large disturbance believed to be typical of those associated with magnetic storms.

Ground-backscatter radars measure the directional and time distribution of ground echoes that return to the radar after propagation over long oblique ionospheric paths. Most of the characteristics of the echoes are determined by irregularities in the ionosphere, but because large volumes of ionosphere are involved, interpretation of the echoes in terms of ionospheric structure is very difficult. Figure 6 shows some

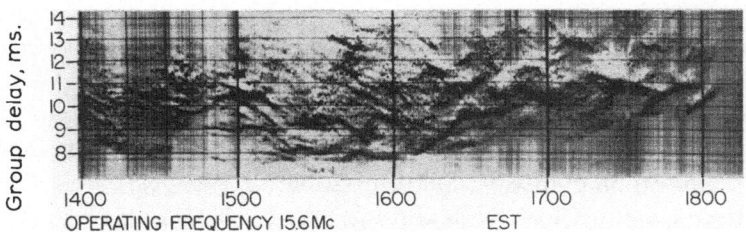

Fig. 6.   Sample 'range-time' ground-backscatter record (after Tveten).

radar returns from a 40° azimuthal sector, plotted in the group-delay-vs.-time plane. The large areas surveyed by a single radar make the technique potentially capable of services that would otherwise require a large number of widely spaced sounders. But before the technique can be of much practical use, the meanings of the great variety of radar signatures of ionospheric irregularities must be understood.

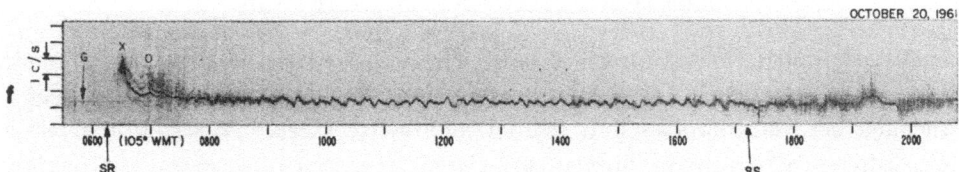

Fig. 7.   Sample frequency–time record made with CW Doppler sounder at vertical incidence.

High-frequency, CW Doppler sounding continuously measures the frequency of stable radio transmissions reflected from the E or F layer, monitoring ionospheric motions by virtue of the Doppler effect. Though somewhat an oversimplification, a frequency-vs.-time record can be thought of as a time history of the vertical velocity of the ionosphere near the radio reflection height. The simplicity of the technique permits continuous, unattended operation, and spaced sounders permit the sizes and velocities of wavelike irregularities to be determined (Georges, 1967). The spectral content of ionospheric motions is preserved by the technique, permitting the dominant periodicities of the motions to be determined. Figure 7 shows a typical frequency–time record for a few hours. Because a fixed frequency is used, daytime and nighttime records represent the motions of different heights in the ionosphere. Multifrequency soundings provide height resolution. In many ways, Doppler sounding is almost the ideal tool for synoptic studies of ionosphere dynamics.

While motions that are clearly wavelike provide opportunities for detailed study, the 'normal' ionosphere seems to be characterized by continuous motions resembling random fluctuations. The fading-correlation and Doppler measurements show this better than other techniques. The idea that even these apparently random motions are caused by atmospheric waves – but by many superimposed waves from many sources – is supported by spectral analysis of these fluctuations. The spectrum of the fluctuations exhibits characteristics that vary with height in the atmosphere in a way that is consistent with the propagation characteristics of atmospheric waves: longer fluctuation periods seem to be characteristic of greater heights.

We have seen that the dynamic characteristic of the ionosphere can affect every characteristic by which a radio wave is specified: its amplitude, phase, frequency, polarization, direction of arrival, and their spatial and temporal derivatives. Fluctuations in these quantities that are almost wavelike or nearly random in character can be expected, but while certain spectral characteristics can be expected on a statistical basis, adequate methods do not yet exist for predicting just what kinds of irregularities will affect a given propagation path at a given time. With a greater understanding of the interaction between the lower atmosphere and the upper atmosphere, between neutral-air motions and ionization motions – both to be discussed in the next section – we can hope for a more accurate characterization of the dynamic ionosphere for telecommunications purposes.

3. Ionospheric Effects of Atmospheric Waves

Acoustic or sound waves form a class of atmospheric waves familiar to everyone. They are longitudinal (compression) waves arising from a nearly adiabatic (lossless) exchange between compression (potential) and inertial (kinetic) energy of a parcel of air. As the wave frequency is lowered, gravitational forces and the density stratification of the atmosphere come into play and give rise to waves with complex and peculiar dispersion characteristics resembling those studied in magnetoionic theory. Furthermore these waves are no longer purely compression waves, but have transverse air motions. Waves in which both gravitational and compressional forces operate are called *acoustic-gravity waves*. When the wave period is longer than about 20 min, the energy exchange is mainly between gravitational potential energy and inertia, and the waves resemble three-dimensional analogs of surface waves on water. These waves are called *internal gravity waves*, the word 'internal' signifying a three-dimensional wave structure.

The propagation characteristics of atmospheric waves are determined largely by the presence of two characteristic frequencies, somewhat analogous to the plasma- and gyrofrequencies in magnetoionic theory: one is the 'acoustic cutoff frequency' (ω_A) and represents a lower frequency limit for waves with acoustic-like properties; the other is the 'Brunt frequency' (ω_B) which forms the upper frequency limit for internal-gravity-like waves. Normally $\omega_A > \omega_B$ so that only 'evanescent' waves (those with no vertical phase variations) exist between ω_A and ω_B. The values of these frequencies are determined by the properties of the atmosphere, and thus vary somewhat with height. Typical values are:

	τ_A	τ_B	(minutes)
Troposphere	4	5	
Ionosphere	14	15	

where τ_A and τ_B represent the periods corresponding to ω_A and ω_B. Further details of the theory of propagation of these atmospheric waves can be found in the works of Hines (1960) and Tolstoy (1963).

Wavelike ionospheric motions have been associated with atmospheric waves mainly by matching the irregularity characteristics (velocities, wavelengths, wavefront orientations) with those of atmospheric waves that would be expected theoretically to exist there. The most notable success has been in identifying some long-period (longer than about 15 min) disturbances with internal atmospheric gravity waves, although there is also some evidence linking shorter-period motions with acoustic waves.

Some properties of internal atmospheric gravity waves that are relevant to ionospheric motions can be briefly summarized as follows:

(a) Sources are believed to lie below the F region; mostly in the troposphere but some probably in the auroral-zone E region.

(b) Phase fronts propagate obliquely downward, with the top of the wavefront tilted forward, when energy propagates upward (as when the source is on the ground).

(c) Under certain asymptotic conditions which are believed to prevail, the forward 'wavefront tilt' tends to increase with wave period, that is, near $\tau = \tau_B$ the front is nearly vertical, while for long-period waves the front is nearly horizontal; these waves are essentially transverse.

(d) Periods must be longer than about 15 min at ionospheric heights.

(e) Damping processes affect short waves more than long ones, and damping increases with height in the atmosphere.

(f) Wave amplitude increases exponentially with height until damping and non-linear processes take over.

(g) Reflection and ducting processes caused by the thermal structure of the atmosphere influence the spectrum of waves reaching the ionosphere from below, and also permit wave propagation over global distances.

(h) Prevailing winds, and the processes by which the waves interact with ionization in the presence of the earth's magnetic field, probably 'select' preferred wavelengths and directions of propagation to be manifested as ionospheric irregularities.

Although many ionospheric motions seem to be attributable to atmospheric waves, the exact mechanism by which such waves are generated and propagate to ionospheric heights is still open to discussion. Atmospheric waves can apparently be generated by a number of natural and artificial sources, so that the presence of a broad spectral 'background' of waves in the atmosphere can be considered normal. Processes that we associate with tropospheric weather, particularly severe weather, are believed to generate sufficient wave energy to reach ionospheric heights. Jet streams, mountain waves and frontal activity have been suggested as likely wave sources, but as yet none has been conclusively identified with ionospheric effects. There is, however, some evidence connecting severe thunderstorm activity with short-period (probably acoustic) waves in the ionosphere (Georges, 1968). Among the artificial sources, nuclear explosions produce long- and short-period ionospheric oscillations that are observable at global distances. TNT explosions, supersonic aircraft and static rocket-engine tests have also been identified with small ionospheric perturbations. Ionospheric perturbations associated with atmospheric waves were identified following the 1964 Alaskan earthquake.

The interaction of atmospheric waves with ionization is believed to take place mainly through momentum exchange in collisions of air molecules with ionized constituents, but other processes, such as alterations of the photochemical balance by the neutral wave, may also be important in some height ranges and for some wave parameters (Hooke, 1968). The way in which atmospheric waves interact with ionization helps to determine which waves have the greatest effects on ionospheric dynamics. In the F region, for example, ionization motions can be considered 'field-dominated', so that waves that have air motions parallel to the earth's magnetic field (for example, internal gravity waves traveling equatorward) are most effective in moving the ionization. In the E region, ionization motions are neither 'collision-dominated' nor 'field-dominated', and the response of the ionization to neutral-air motions is complicated by the generation of electric currents (Ratcliffe, 1959).

Very long atmospheric waves called atmospheric tides have been previously mentioned in connection with sporadic-E layers. They are wave motions resembling internal gravity waves, but have periods that are submultiples of the earth's rotation period, and a more complicated behavior because of the influence of the earth's rotation. They are excited mainly by thermal energy supplied by the sun.

4. Computer Simulation

The various radio observations discussed in Section 2 form only half the effort to understand ionospheric dynamics and its radio effects. The other half consists of ionosphere modeling and experiment simulation. In theory there ought to be continuous interaction between actual observations and those predicted by theoretical models, but in practice it seldom works out that way. At present a severe lack of cross-fertilization exists among the various radio observations, and between the observers and the theorists, partly because of the difficulty of expressing the results of each observation in terms of the common denominator – the ionospheric motions themselves. Digital-computer radio ray tracing seems to provide a key for unraveling the tangle of apparently unrelated observations, and this approach is now being pursued actively at the ESSA Research Laboratories.

The behavior of radio waves in an ionosphere with an irregular structure is far too complicated to be treated in an analytical way; only grossly oversimplified model irregularities lend themselves to such techniques. On the other hand, ray tracing, in its present state of development, permits a computer simulation of almost any radio measurement one could make, using practically any realistic model ionosphere. (The exceptions are phenomena that cannot be accurately represented by ray theory, such as caustics and partial reflections.)

The general procedure is to construct model ionospheric disturbances that are both theoretically sound and consistent with the more reliable radio observations, then to construct, using ray tracing, synthetic 'data' as they would be recorded in the observation in question. Synthetic data are compared with samples of actual data, and, with the auxiliary information that ray-tracing programs provide, detailed processes of the interaction of the radio rays with the irregularities are determined. Wave and irregularity parameters are varied to determine the effects of changes in operating parameters and changes in irregularity properties, and also to find out what features of the irregularities are most effective in altering the properties of the radio wave.

When several observational techniques are simulated using the same ionospheric model, apparently diverse observations can be shown to be simply different manifestations of the same ionospheric phenomenon. For example, this approach has been used to demonstrate the equivalence of a certain kind of TID signature on CW Doppler records with another signature on h', t records. This permits the large

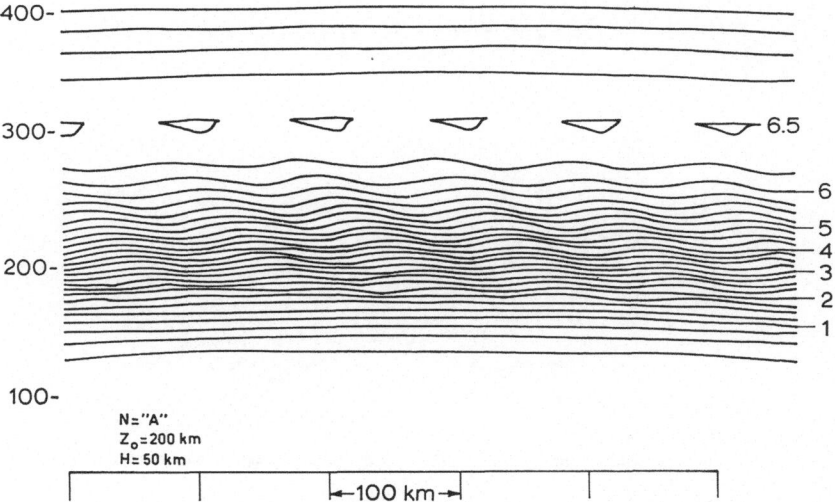

Fig. 8. Electron density contours for a wavelike model irregularity used in ray-tracing-simulation studies. The scale is 100 km/div, and the contours are plasma frequency in MHz.

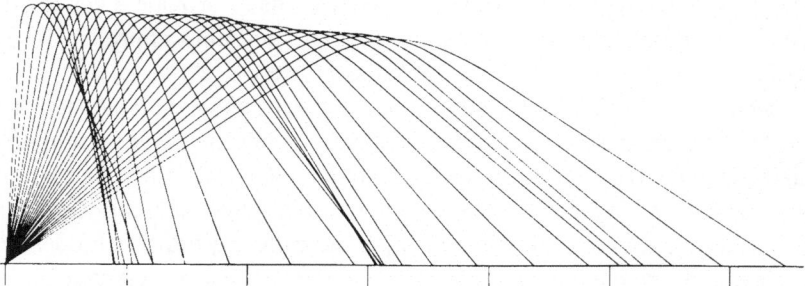

Fig. 9.   Some 4 MHz raypaths traced through the model of the preceding figure.

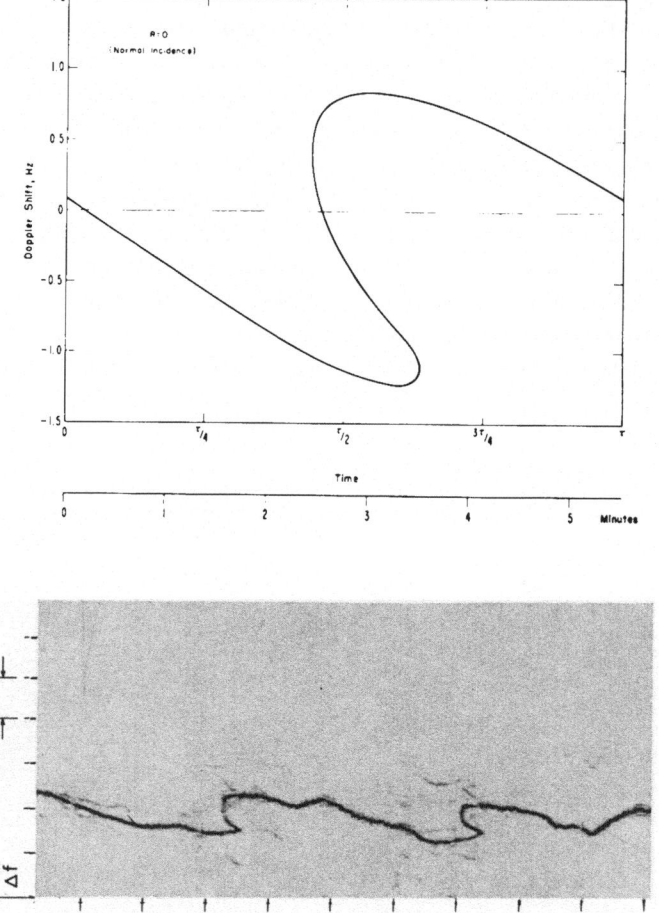

Fig. 10.   A simulated frequency–time record using the model of Figure 8, compared to the actual
Doppler signatures of two TIDs.

amount of TID data collected with fixed-frequency pulse sounders to be related to certain kinds of Doppler signatures.

Preliminary simulation results have been obtained for HF ground-backscatter and direction-of-arrival measurements, in addition to the two techniques just mentioned, and the results look promising. Recent advances in ray tracing have permitted a simplified calculation of Doppler shift using a single ray, and the use of arbitrary,

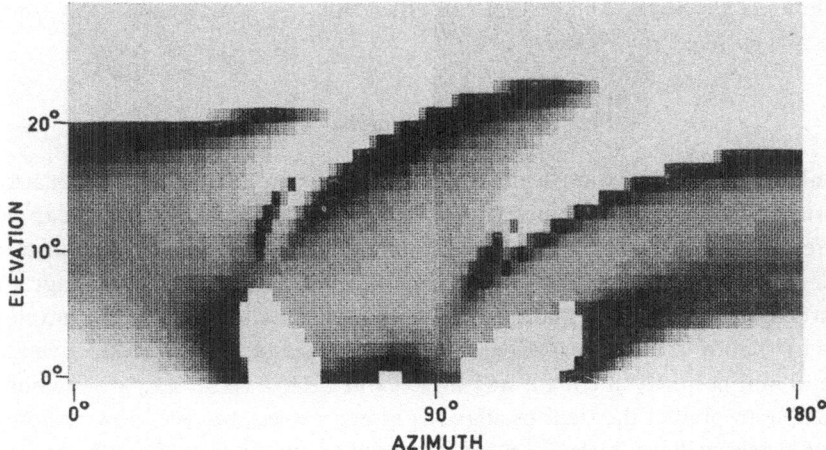

Fig. 11. A simulated ground-backscatter record in the plane defining the direction of transmission. The black areas are directions of strongest echoes, while the white areas denote rays that penetrate the ionosphere or miss the earth.

time-varying three-dimensional ionosphere models. Formulations of the field-strength expressions for the three-dimensional case have been carried out, permitting the calculation of backscatter amplitude in three dimensions. Amplitude calculations are made by tracing several adjacent rays and applying conservation of energy to the resulting flux tubes. To keep the results and interpretation as simple as possible, the effects of the earth's magnetic field (anisotropy) and electron collisions (losses) are neglected. Both exert relatively minor influences when the radio frequency is much greater than the gyrofrequency (about 1.5 MHz at midlatitudes).

The electron density models used in these studies are all time-varying, three-dimensional wavelike irregularities superimposed on Chapman-layer ambient distributions. Figure 8 shows an example of a wavelike model irregularity (a North–South section) that is believed to be an idealization of a type of disturbance commonly seen in the daytime F layer, and Figure 9 shows some 4 MHz raypaths calculated for this model.

Figure 10 shows a simulated frequency–time record compared with an actual Doppler record showing two TIDs.

Figure 11 is an example of the simulated ground-backscatter intensity (black strongest) that would be measured by a pencil-beam radar scanning from north to south in azimuth (φ) and from horizontal to penetrating takeoff angles (β). The wave frequency is 15 MHz and the ionosphere model is a single Chapman layer with a wavelike irregularity similar to that in Figure 8, but with a horizontal wavelength of 1000 km, traveling Southward. This is an example of a simulation for which no corresponding actual measurements yet exist.

In such a synthesis approach, the question of uniqueness inevitably arises. While it is true that there is no guarantee that agreement between synthesized data and real data means the model is all right, one can feel fairly confident about a model that is both theoretically sound and predicts the effects that are actually observed. One could hardly ask more of any model.

5. Conclusion

The only remaining question seems to be – what do we do with all the answers that the foregoing research will someday provide us with? Can we really hope to describe and predict the ionosphere's small-scale, time-varying structure as completely as we now describe its long-term, global-scale behavior? Clearly not. The analogy with the sea surface is again appropriate here. There is no deterministic description of the detailed structure of all the waves and motions on the ocean surface; yet we know in a general and perfectly adequate way how it will behave under any given conditions: we are able to predict the tidal oscillations at every seashore; we know roughly what kind of waves will be kicked up by a hurricane, and can predict tsunami waves generated by seismic activity; and we can predict with reasonable accuracy the wavelengths, amplitudes, speed and direction of the waves on the relatively calm 'normal' surface, given the surface wind conditions. This is what we would like to be able to do for the ionosphere. Knowing all the factors that influence ionospheric motions, we should be able to characterize the kinds of wave motions that will be seen under various normal and disturbed conditions. But since the ionosphere is not quite as observable as the ocean surface, we obviously have to be a little more clever about it.

References

Aarons, J. (ed).: 1963, *Radio Astronomical and Satellite Studies of the Atmosphere*, North-Holland Publ. Co., Amsterdam.

Booker, H. G.: 1958, 'The Use of Radio Stars to Study Irregular Refraction of Radio Waves in the Ionosphere', *Proc IRE* **46**, 298–314.

Bramley, E. N.: 1953, 'Direction-Finding Studies of Large-Scale Ionospheric Irregularities', *Proc. Roy. Soc. London* **220**, 39–61.

Briggs, B. H. and Spencer, M.: 1954, 'Horizontal Movements in the Ionosphere', *Rep. Progr. Phys.* **17**, 245–280.

Georges, T. M.: 1967, 'Ionospheric Effects of Atmospheric Waves', ESSA Tech. Rept. IE R-57/ITSA-54, U.S. Govt. Printing Off., Washington, D.C.

Georges, T. M.: 1968, 'Short-Period Ionospheric Oscillations associated with Severe Weather', in *Acoustic-Gravity Waves in the Atmosphere – Symposium Proceedings*, U.S. Govt. Printing Off., Washington, D.C.

Herman, J. R.: 1966, 'Spread-*F* and Ionospheric *F*-Region Irregularities', *Rev. Geophys.* **4**, 255–299.

Hines, C. O.: 1960, 'Internal Atmospheric Gravity Waves at Ionospheric Heights', *Can. J. Phys.* **38**, 1441–1481.

Hooke, W. H.: 1968, 'Ionospheric Irregularities produced by Internal Atmospheric Gravity Waves', *J. Atmos. Terrest. Phys.* **30**, 795–823.

Munro, G. H.: 1958, 'Travelling Ionospheric Disturbances in the *F*-Region, *Aust. J. Phys.* **11**, 91–112.

Ratcliffe, J. A.: 1959, 'Ionizations and Drifts in the Ionosphere', *J. Geophys. Res.* **64**, 2102–2111.

Rishbeth, H.: 1967, 'A Review of Ionospheric *F*-Region Theory', *Proc. IEEE* **55**, 16–35.

Rishbeth, H. and Garriott, O. K.: 1964, 'Introduction to the Ionosphere and Geomagnetism', Stanford Electronics Lab., Tech. Rept. No. 8.

Smith, E. K. and Matsushita, S.: 1962, *Ionospheric Sporadic E*, Pergamon, New York.

Thome, G. D.: 1964, 'Incoherent Scatter Observations of Travelling Ionospheric Disturbances', *J. Geophys. Res.* **69**, 4047–4049.

Tolstoy, I.: 1963, 'The Theory of Waves in Stratified Fluids including the Effects of Gravity and Rotation', *Rev. Mod. Phys.* **35**, 207–230.

SOME CHARACTERISTICS OF NON-TWO-STREAM
IRREGULARITIES IN THE EQUATORIAL ELECTROJET*

BEN B. BALSLEY

Jicamarca Radar Observatory**, Lima, Peru

Abstract. The analysis of VHF radar echoes from electron density irregularities in the equatorial electrojet indicate that there are two distinct types of irregularities in the region. One of these has previously been shown to be generated by the two-stream instability mechanism and consists of planar irregularities which travel at the ion-acoustic velocity. The other type of irregularity moves at about the electron drift velocity, and can exist when the electron drift velocity is insufficient to produce the two-stream type irregularities. Drift observations of this second type of irregularity show that the electrojet current reverses during nighttime periods, the electron drift velocity being comparable to that during the day. Additional observations show that these irregularities exist in large patches which move with the same approximate velocity and retain their identity for many seconds.

1. Introduction

The spectra of VHF radar echoes from electron density irregularities in the equatorial electrojet show that the irregularities are produced by the two-stream instability (Farley, 1963) when the electron drift velocity is large enough. These two-stream irregularities are plane-wave structures which travel at the ion-acoustic velocity (approximately 360 m/sec). However, when the current density is low, or when the echoing region is examined by other than spectral techniques (albeit at the same frequency), not all of the observed irregularity characteristics can be accounted for by the two-stream mechanism. Specifically:

(a) When the current density is low, the echoes are Doppler-shifted by amounts much less than that predicted by the two-stream theory (Farley, 1963; Balsley, 1966; Dougherty and Farley, 1967; Cohen and Bowles, 1967). At times, such returns are stronger than those from the two-stream irregularities, particularly when the radar antenna is directed vertically.

(b) Large 'patches' of the small (~ 3 m) irregularities typically detectable at VHF are observed in the region (Balsley, 1965). These patches have dimensions of about 1 km and lifetimes of many seconds, and move at appreciable velocities relative to a ground-based observer.

(c) Although midday echoes normally come from the entire vertical extent of the echoing region, a marked splitting (bifurcation) of the echoing region occurs at times well away from midday, when the electrojet intensity is considerably less than its normal midday value (Bowles and Cohen, 1962; Balsley, 1965).

The above observations indicate that there may be two separate types of irregularities in the electrojet which are produced by separate mechanisms. A qualitative theory

---

* This is a reprint of a paper published in *J. Geophys. Res.* **74** (1969) 2333.
** A cooperative project of the Aeronomy Laboratory, ESSA Research Laboratory, Boulder, Colorado and the Instituto Geofisico del Peru, Lima, Peru.

D'Angelo (ed.), Low-Frequency Waves and Irregularities in the Ionosphere. All rights reserved

to account for the spectral characteristics of the irregularities not generated directly by the two-stream mechanism (referred to here as non-two-stream irregularities) has been published by Dougherty and Farley (1967) (cf. also a companion paper by Cohen and Bowles, 1967). Dougherty and Farley propose that the waves which are generated by the two-stream instability will grow until they are limited by non-linear effects, and postulate that these effects will cause coupling between waves, which will, in turn, generate new waves that cannot be accounted for in the linear theory. Although characteristics such as the large patches and the bifurcation cannot be accounted for by their theory, they are able to account for most of the observed spectral characteristics by vectorially combining plane waves with slightly different direction vectors.

This paper is the result of a series of investigations into the properties of the non-two-stream irregularities. All of the investigations were made with the VHF (49.920 MHz) radar system at the Jicamarca Radar Observatory near Lima, Peru. Geographic co-ordinates of the station are 11° 57.0' South and 76° 52.0' West. The magnetic dip is about 2° North. The experiments were performed between September, 1966, and November, 1967.

2. Presentation and Preliminary Analysis of Data

A. GENERAL REMARKS

1. *Methods of Analysis*

The majority of the electrojet irregularity characteristics reported on here have been obtained from the spectral analysis of echoes returned from the irregularities. This procedure has been used extensively in the past at this Observatory, and is discussed in detail in previous papers (see, for example, Cohen and Bowles, 1967).

In the case of the electrojet, analysis of the spectral content of the echoes returned with different time delays after the transmitted pulse corresponds to a spectral analysis at different antenna zenith angles. This is because of the relatively thin ($\simeq 10$ km) echoing region and the aspect sensitivity of the irregularities (Bowles *et al.*, 1963) which limits the observable echoing region to a narrow band in the magnetic East–West/vertical plane. Hence, provided that the antenna system is directed to either side (magnetic eastward or westward) of the vertical, the signals returned at greater time delays correspond to echoes arriving at greater zenith angles from this (East or West) direction.

In addition to the spectral analysis, some data have been obtained by the range-time-intensity (RTI) method which has also been used previously at this station (Balsley, 1965). In this type of analysis, the echo intensity is used to modulate the intensity (not the amplitude) of an oscilloscope trace which has been calibrated in time delay after the transmitted pulse. This presentation is photographed by continuously drawing a film strip across the cathode ray tube at right angles to the trace. The resulting film strip contains information on the echo strength as a function of time and range (or height if the antenna has a narrow beam and is directed vertically).

2. Demonstration of the Existence of the Non-Two-Stream Irregularities

A typical example of echo spectra taken at a relatively oblique antenna angle during periods of high electrojet intensity is shown for reference in Figure 1(a). The 125 Hz displacement (approximate) of the spectra peak from the transmitted frequency corresponds to the Doppler shift of a 50 MHz signal returned from irregularities which are traveling at the ion-acoustic velocity. This shift, coupled with the fact that it is virtually independent of antenna elevation angle, shows that the irregularities are plane wave structures which are generated by the two-stream instability (Farley, 1963).

Spectra which depart from this 'classical' example of the two-stream spectra are shown in the following three figures. Figure 1(b) was taken at almost the same time as 1(a) but at a less oblique angle; Figure 1(c) was taken at the same elevation angle as 1(a) but later in the afternoon when the electrojet intensity (determined from

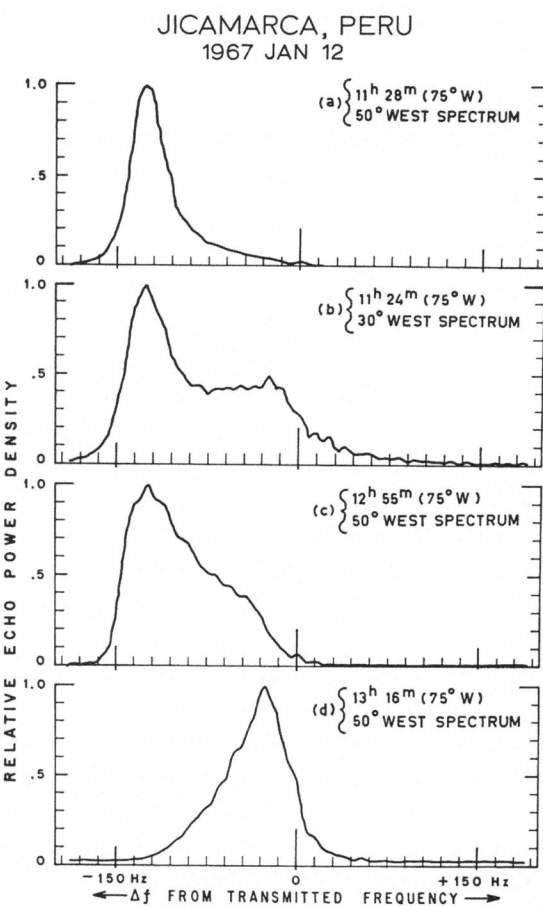

Fig. 1.   Spectra taken with the antenna directed toward the West, showing the presence of the non-two-stream irregularities. (a) 1128 hours with $\theta = 50$ degrees, (b) 1124 hours with $\theta = 30$ degrees, (c) 1255 hours with $\theta = 50$ degrees, (d) 1316 hours with $\theta = 50$ degrees.

concurrent magnetometer records) had diminished from its maximum midday level. Although both of these spectra show evidence of the two-stream echo, it is apparent that an appreciable portion of the power was returned from irregularities which were moving (relative to the observer) at velocities considerably less than the ion-acoustic velocity.

The final spectrum of this series (Figure 1(d)) was taken at a time when the magnetometer indicated that the electrojet current was quite weak. In this spectrum, there is no evidence of the two-stream echoes, but only a weaker echo whose frequency shifts are much less than those of the two-stream type. (The strength of the echo is not shown on these spectra since they have been plotted on a normalized format.)

These last three spectra, then, show that, under some conditions, not all of the echoes are returned from the two-stream irregularities. This observation implies that at least one other type of irregularity must exist in the region, at least under some conditions.

3. Definition of Irregularity Types

For the sake of convenience, the electron density irregularities which are generated directly by the two-stream instability will be classified in this report as Type I irregularities; echoes from these irregularities will be classified as Type I echoes. Irregularities and echoes of the non-two-stream type will be classified as Type II.

B. TYPE II IRREGULARITY MOTION DEDUCED FROM SPECTRAL DATA

1. Demonstration that the Mean Motion is Horizontal

The spectra shown in Figure 2 were taken simultaneously at three separate antenna zenith angles by the methods outlined in the previous section. From these spectra, it is apparent that the mean Doppler shift (defined by the dashed vertical line which separates the spectrum into two equal parts) is greater for signals returned at more oblique angles.

A plot of the mean Doppler shift, in terms of an observed radial drift velocity V_{obs}, versus antenna zenith angle for 10 such simultaneous spectra is shown in the same figure. The quantity V_{obs} is defined by

$$V_{obs} \equiv c\Delta f/(2f) \text{ m/sec},\qquad (1)$$

where c is the velocity of light (in vacuo) expressed in m/sec, f is the transmitted frequency in Hz, and Δf is the mean Doppler shift (defined above) in Hz.

The three continuous curves which appear in the same plot represent the Doppler shift (in terms of V_{obs}) as a function of zenith angle which would be expected if the echoes were returned from irregularities drifting Westward (at velocities $V_d = 225$ m/sec, 250 m/sec and 275 m/sec) at a height of 107 km. The agreement between the data points and the theoretical curve for V_d 250 m/sec is seen to be quite good.

The above example is typical of a number of analyses made during the experimental period. In every case, the Type II irregularities were found to be drifting horizontally.

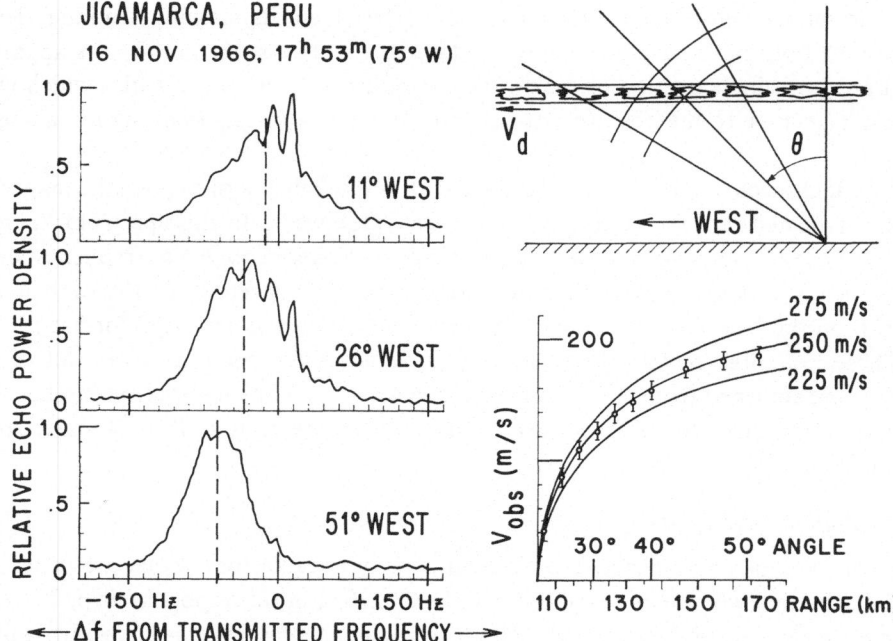

Fig. 2. Showing three Type II spectra taken simultaneously at different antenna zenith angles. Mean Doppler shift shown by the dashed lines. The results of the complete analysis appears in the lower right-hand section and demonstrate that the irregularities are drifting horizontally at about 250 m/sec. The geometry of the experiment appears in the upper right-hand part of the figure.

Drift velocities ranged from about 50 m/sec to at least 360 m/sec. For velocities below 50 m/sec, the echoes became very weak and contamination from background noise rendered analysis impossible. For velocities above 360 m/sec, the spectra became contaminated by the Type I echoes, making analysis quite difficult. In some cases it was possible to infer velocities greater than 360 m/sec by examining only those spectra taken near the vertical, where the Type I contamination was least. The results again showed evidence of horizontal drifts, but because the complete analysis over a wide range of angles was not made, the significance of the measurements was somewhat in question.

Within the range of experimental error, it is possible that the irregularity drift velocity has a small vertical component. Indeed, Sugiura and Cain (1966) show that the magnitude ratio of the horizontal and vertical drift velocities of *electrons* is in proportion to the ratio of the Hall and Pedersen conductivities (about 30:1 at the height of maximum electron drift velocity). Insofar as the Type II irregularities move in concert with the electrons (cf. following sections), one would expect that the irregularities partake of the vertical drift.

2. *Type II Irregularity Drift Direction compared with the Direction of Electron Drift*

Nighttime spectral records indicate that the Type II irregularities are moving eastward in contrast to their daytime westward motion (Balsley, 1966). A similar behavior is predicted for the electrojet electrons from dynamo theory, implying a close relation between the Type II irregularity motion and the electron flow. Although the daytime westward flow of both the electrons and the irregularities is established, only the eastward (nighttime) motion of the Type II irregularities has been observed; proof of the actual eastward electron flow during nighttime periods is needed. Observational evidence for this Eastward flow is shown by the nighttime spectrum in Figure 3, taken with the antenna directed westward. The pronounced peak due to the Type I irregularities at about 120 Hz above the transmitted frequency indicates that the Type I irregularity phase velocity is toward the East. Since there is strong theoretical support that these instabilities travel in the general direction of the electron drift, this must be construed as a definite indication of the current reversal. (Although only one example is shown here, this feature is quite common on nighttime spectra.)

Type II irregularity drift direction during morning and evening periods is shown in Figure 4 for all instances during the period September 1966 to February 1967. Both the westward to eastward drift reversal of the electrons during the evening hours,

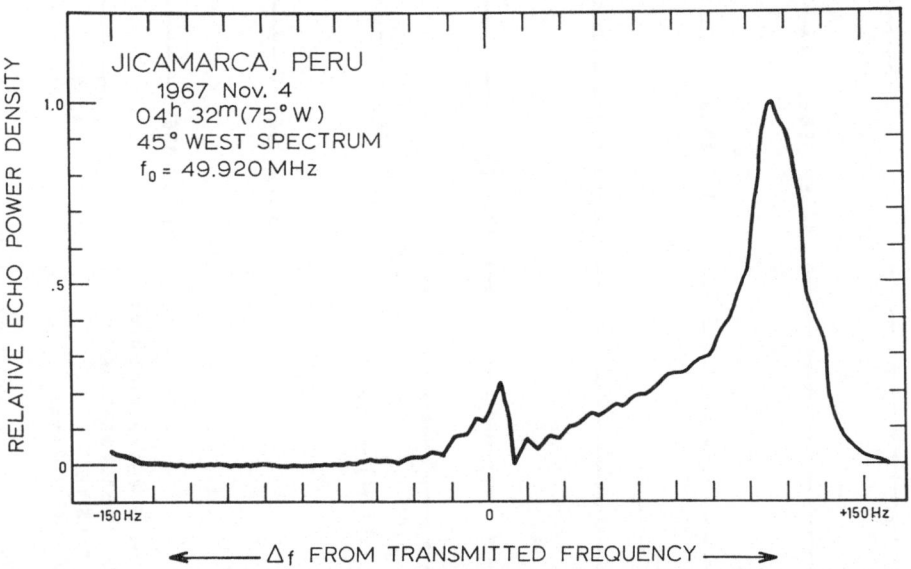

Fig. 3.   Spectrum taken at night with the antenna directed toward the West. The positive Doppler shift of the two-stream echoes indicates that the electrons are moving eastward in contrast to the normal westward flow during the day.

and the converse during the morning hours, are seen in this figure. Although the data are not presented here, the eastward nighttime motion of the electron continues throughout the night. The reversal times appear to center around 0650 hours for the morning reversal and 2030 hours for the evening. These times should be compared with the theoretically predicted times of about 0610 hours and 1900 hours inferred from Maeda and Kato (1967).

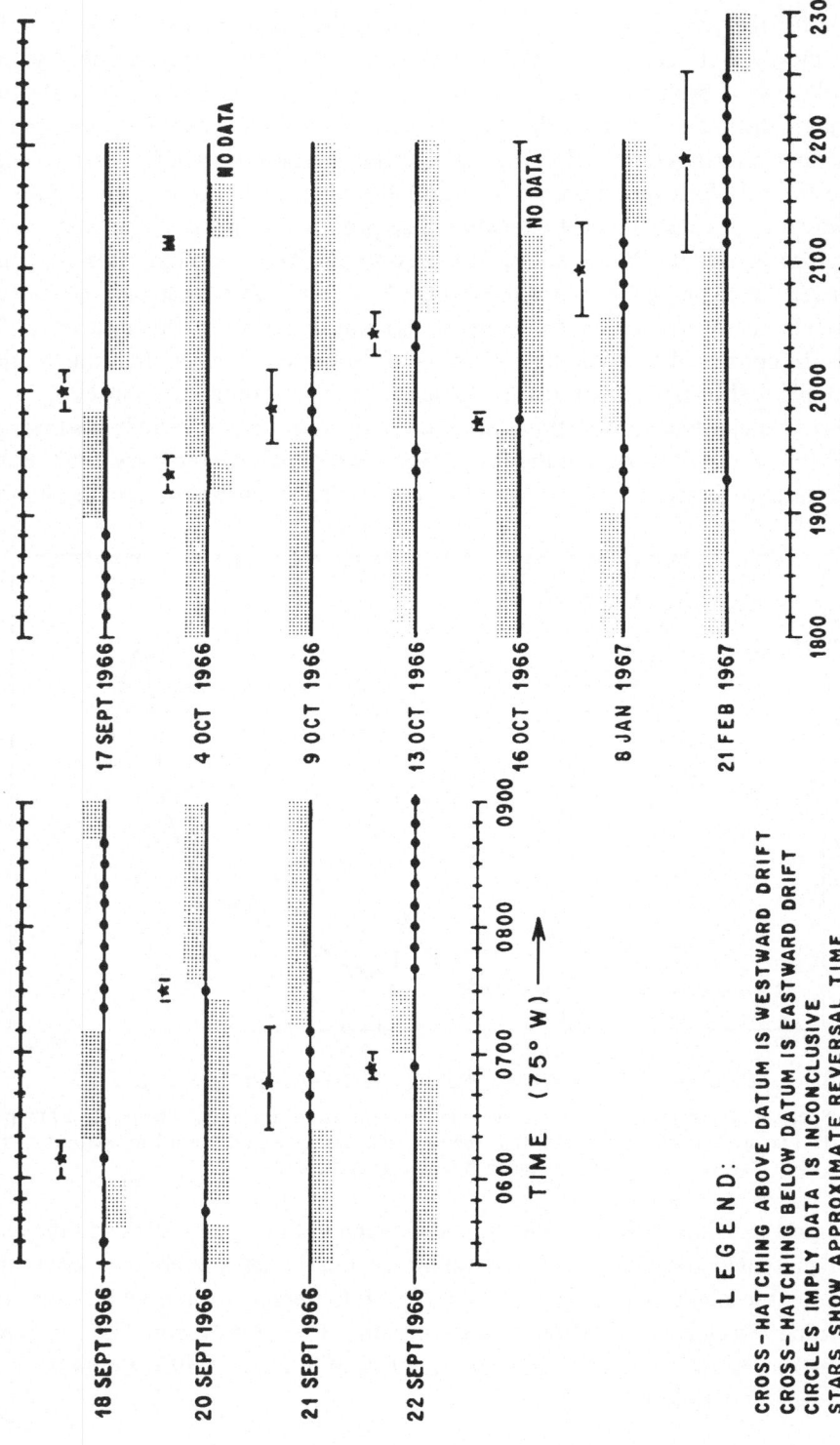

Fig. 4.   Showing Type II irregularity drift *direction* during morning and evening periods. The drift reversal times are indicated roughly by the stars.

3. *Demonstration that Type II Drift Velocities are Close to the True Electron Drift Velocities*

Two comparisons can be made between the magnitudes of the Type II irregularity drift velocity and the electron drift velocity. The first of these is that the Type II irregularity drift reverses at the time that one would expect a reversal of the electron drift, so that the electron drift velocity and the Type II irregularity drift velocity are zero at nearly the same time (see the previous section).

Secondly, the Type I (two-stream) irregularities, which should be present only when the electron drift exceeds the ion-acoustic velocity, appear only when the velocity of the Type II irregularities exceeds this value (360 m/sec). A typical example of the second observation is presented in Figure 5, where the average drift velocity of the

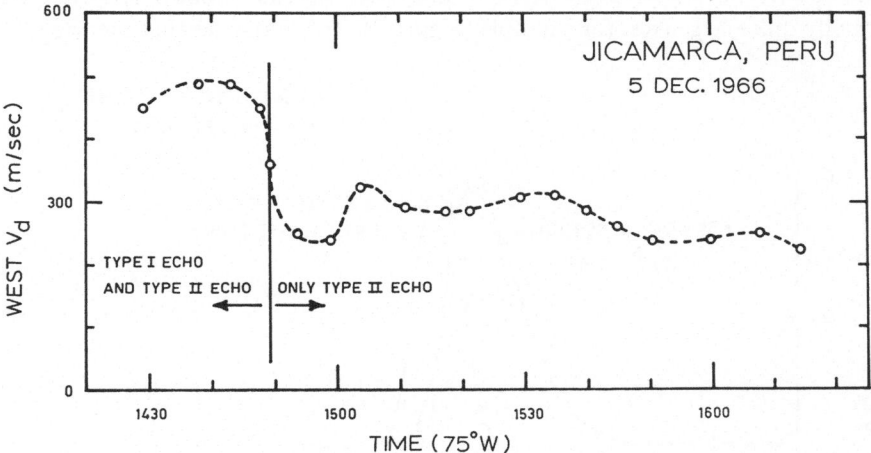

Fig. 5. Drift analysis of Type II echoes, showing that the Type I irregularity disappears for drift velocities less than about 360 m/sec.

Type II irregularities is shown as a function of time. Evidence of both Type I and Type II echoes was present on the spectra until 1450 hours local time (the spectra were taken at angles close enough to the vertical that the two types of echoes could be resolved). After 1450 hours, no trace of the Type I echoes was seen. The Type II drift velocity at this time was about 360 m/sec.

Two points, then, may be fixed for the drift velocities of the Type II irregularities. At these two points the irregularity drift velocity is seen to be in agreement with the mean drift velocity of the electrons.

Within these limits, it would be desirable to determine the relationship between the two. The only possible way of doing so at present is by means of concurrent magneto-meter records, since the magnetometer affords a sensitive measure of the electrojet current, and since the current is proportional to the electron velocity. Indeed, a qualitative comparison shows that a decrease (or increase) in the Type II drift velocity corresponds to a decrease (or increase) in the magnetic intensity. Unfortunately, a more definite relationship cannot be established without a number of assumptions because the magnetometer reading is also proportional to the electron density as well as the cross-sectional area of the electrojet region. Changes in either of these two parameters will, therefore, result in a change in the observed magnetic intensity. Ionospheric currents and magnetic fields outside of the electrojet will also affect the magnetometer readings.

In spite of these problems, the simplest assumption (consistent with the qualitative observation discussed in the previous section) is that the Type II irregularity drift velocity is linearly related to the electron drift velocity within the limits stated above.

4. *Non-Horizontal Motions of Type II Irregularities*

Although the mean Doppler shift of the Type II echoes determines the horizontal drift velocity of the irregularities, the frequency spread about this mean value is normally quite large (see, for example, Figure 2). This indicates that the irregularities

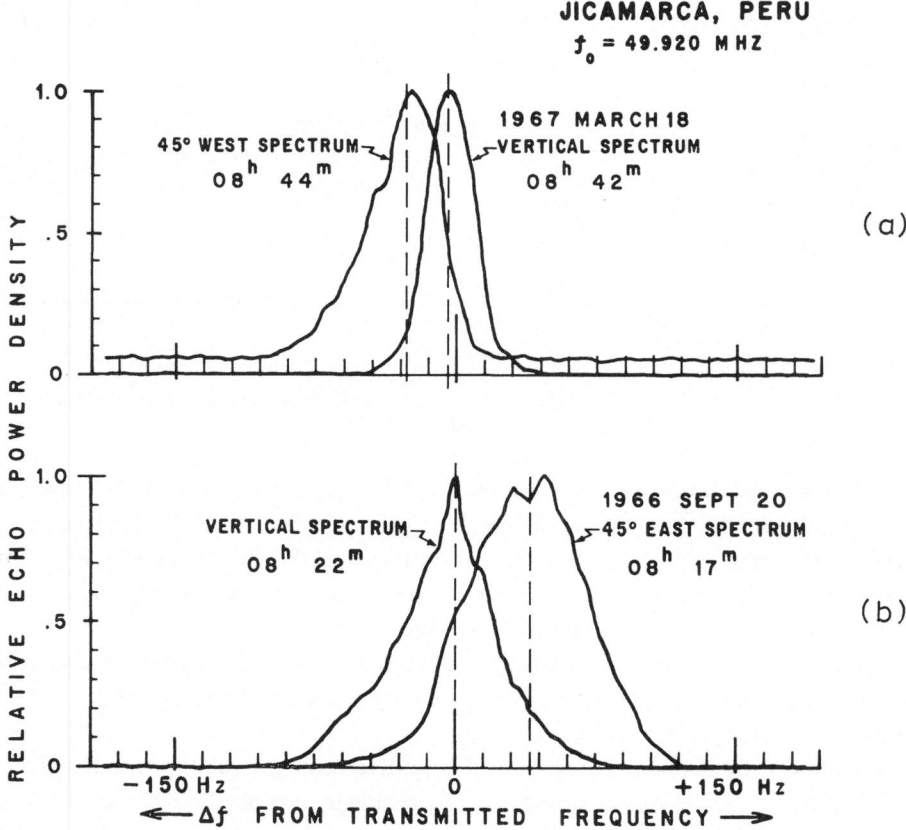

Fig. 6. Vertical and oblique spectra comparing vertical and horizontal motions of the Type II irregularities. The lack of an appreciable mean Doppler shift on the vertical records shows the lack of mean vertical motion. The spread of Doppler shifts on both vertical and oblique spectra attest the rough 'isotropy' of the 'random' velocity components.

are not just drifting horizontally at some particular velocity, but rather that the motion is more complex.

A further insight into this phenomenon is gained by examining the spectra of Type II echoes which were obtained by the Observatory's large incoherent-scatter antenna. Since the antenna beamwidth is narrow ($\simeq 1°$) and directed vertically, any frequency-shifted echoes which may appear on this type of spectrum must be interpreted as resulting from the vertical motion of the Type II irregularities.

A spectrum obtained using the large antenna is shown in Figure 6(a), along with one taken almost simultaneously with the obliquely-directed steerable antenna. A similar set of spectra, obtained when the Type II irregularity drift velocity was somewhat higher, is shown in Figure 6(b). The mean Doppler shifts on all spectra are indicated by the dashed lines.

Examination of the vertically-obtained spectra shows that the Type II irregularities have pronounced vertical motions even though the mean Doppler shifts on the same records show little, if any, indication of vertical drift. Moreover, in both cases the spread of Doppler shifts on the vertical and oblique spectra are comparable (although the oblique-spectral spread appears to be some 50% greater at the half-power points, this discrepancy decreases at the lower power levels). This similarity of Doppler spreading on vertical and oblique spectra implies a rough isotropy of the East–West/vertical velocity distribution of the 'random' velocity components (i.e., those components which contribute to the spreading about the mean drift velocity). If the random components were confined to the vertical direction, then the oblique spectra would be less spread. Conversely, if they were confined to the horizontal, then the vertical spectra would appear much narrower.

Also demonstrated in Figure 6 is a roughly linear relationship between the mean Doppler shift and the width of the spectrum. The difference of about 50% in the mean Doppler shifts between the first and second oblique spectra is reflected as an equivalent difference in the spectral widths.

C. RELATIVE POWER OF TYPE I AND TYPE II ECHOES

The ordinate (relative echo-power density) of a typical electrojet spectrum is plotted on a normalized format, because of the wide range (> 40 db) of average power levels which may be encountered during an experiment. The relative power levels *between* spectra may be obtained by 'unnormalizing' the individual records. This can be done by taking account of the first point of the autocorrelation function, ($\varphi_{\tau=0}$), from which the spectra were derived, since this value is a measure of the total power received during a given spectral sample.

Furthermore, the relative contributions of Type I and Type II echoes on a *particular* spectrum may be determined by assuming that the Type I echoes produce a symmetrical spectrum about the ion-acoustic peak. This is reasonable, since it is true when the Type I echoes are the only ones apparent on a spectrum. With this assumption, the ratio of the power returned from Type I and Type II irregularities is the ratio of their respective spectral areas: the area under the spectral curve due to the Type I echoes being the symmetrical curve discussed above; the contribution of the Type II echoes accounting for the remainder of the area under the total curve.

The contributions of the two irregularity types as a function of total (relative) power may be obtained by combining the two techniques just described; the accuracy of this type of qualitative measurement is clearly best when the spectral areas due to the two types of echoes are not too disproportionate.

A series of three spectra from an experiment of this type is shown in Figure 7.

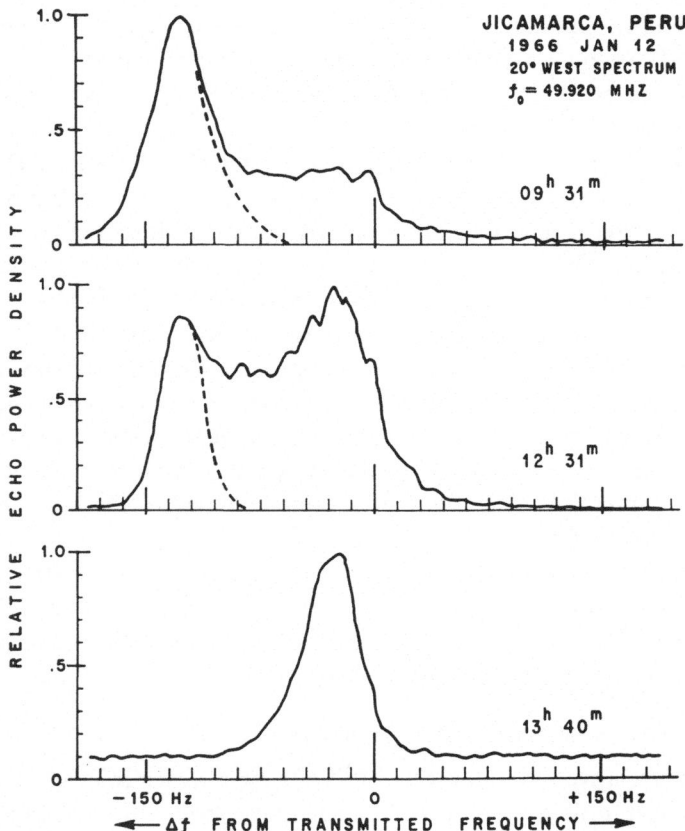

Fig. 7.   Sample spectra to establish the relative power relation between Type I and Type II irregularities. The dashed curves separate the assumed contributions of the two irregularity types.

Note that in every case the maximum power density is plotted at the same height, although there was a difference (not indicated) of more than 17 db between the total echo power received on the first (strongest) and last (weakest) of the three spectra. Also, the assumption of a symmetrical Type I echo spectrum is shown by the dashed lines on the first two of the spectra (note that the last spectrum shows no evidence of the Type I echo).

A graph of the complete results of this analysis is shown in Figure 8. The data were taken at a relatively small zenith angle (20°) so that the two spectral areas were comparable even when the Type I echoes were predominant at the more oblique angles. (For a discussion of the zenith angle dependence of the Type I echo cross-section, see Bowles *et al.*, 1963.) These observations may be made from this figure:

(1) Type II irregularities can exist independently of the Type I.

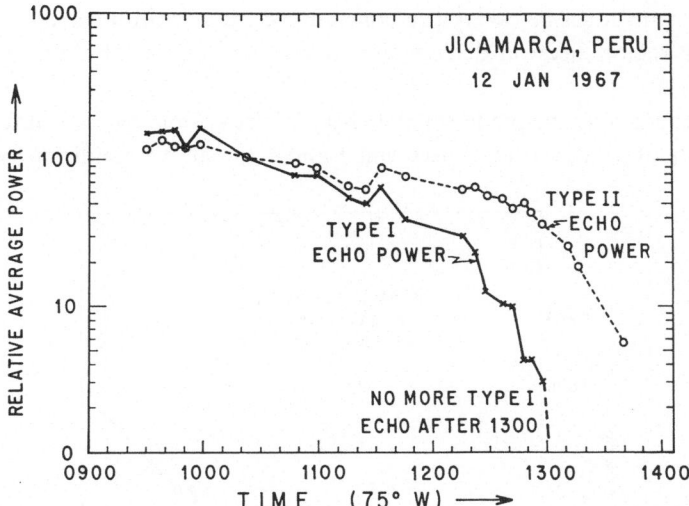

Fig. 8.   Relative strength of Type I and Type II echoes vs. time. Antenna directed Westward at $\theta = 20°$.

(2) An increase (or decrease) in the echo-power of either type of echo is accompanied by a similar change in the other.

(3) When the Type I echoes are quite weak, the rate of change of Type I echo-power, relative to the total echo-power, is much greater than that of the Type II echoes.

The first of these observations shows that the Type II irregularities cannot arise from the Type I irregularities, as postulated by Dougherty and Farley (1967), since they can be present on electrojet spectra even when the Type I irregularities are not.

The last two observations may be used to explain one puzzling aspect of the electrojet spectra: oblique spectra taken during periods of maximum electron drift velocity (when the echoes are quite strong) show very little evidence of Type II echoes, even though weaker signals, at the same angle, show more evidence of the Type II echoes (cf. Figure 1). This would seem to indicate that the strong Type I echoes tend to preclude the presence of the Type II echoes. Observations 2 and 3, however, show that this apparent decrease in Type II echo signal with increasing echo-power is a result of plotting the spectra on the normalized format; both types of signals increase as the total power increases, but the Type II increases much more slowly than the Type I.

D. TYPE II ECHO POWER AS A FUNCTION OF DRIFT VELOCITY

In order to determine a relationship between the relative power returned from the Type II irregularities and their mean drift velocity, one must obtain a series of un-normalized spectra (cf. previous section) under the following conditions:

(1) A sufficient number of spectra must be obtained with mean drift velocities which are fairly evenly distributed between the minimum measurable velocity (about 50 m/sec) and the maximum useable velocity (360 m/sec).

(2) Radar system gain changes (transmitter power variations and receiver attenuator settings) must be taken into account.

(3) All spectra must be taken at a constant antenna zenith angle.

(4) The E-region electron density should not vary appreciably during the experimental period.

(5) Allowance must be made for the effect of the cosmic background noise.

The results of an experiment performed under the above conditions are shown in

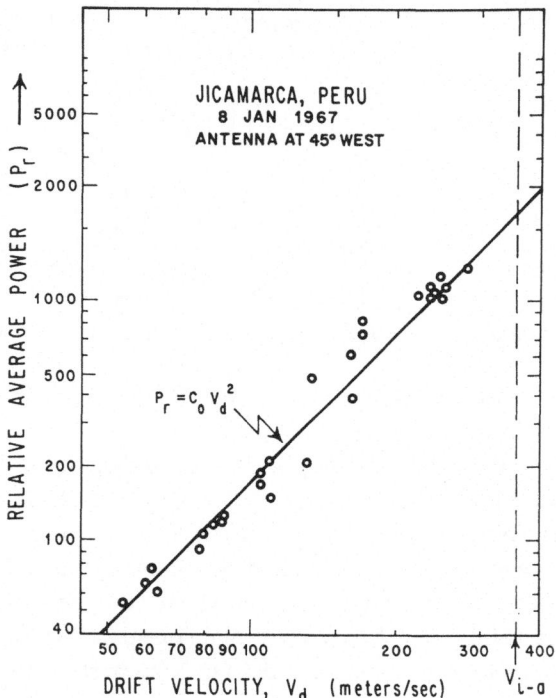

Fig. 9. Type II drift velocity vs. relative average power, showing that the Type II echo power is proportional to the square of the drift velocity.

Figure 9. The data points, which encompass drift velocities between about 50 m/sec and 300 m/sec, are well grouped about a straight line on this graph. Since the coordinates are plotted on a logarithmic scale, the slope of the line gives the power law relationship between the variables. In this case, the relation is given by

$$P_r = C_0 V_d^2, \tag{2}$$

where P_r is the relative average power returned from the Type II irregularities, V_d is the irregularity drift velocity, and C_0 is an arbitrary constant.

E. A COMPARISON BETWEEN THE APPARENT DRIFT VELOCITY OF THE 'PATCHES' AND THE
DRIFT VELOCITY OF THE TYPE II IRREGULARITIES

An apparent drift velocity of the large 'patches' of electrojet irregularities may be
inferred from the slope of the lined structures observed on range-time-intensity (RTI)
film strips of the electrojet echoes (Balsley, 1965). In order to make a comparison
between this velocity and the drift velocity of the Type II irregularities, oblique spectra
were obtained concurrently with the RTI film-strip data during periods when the
magnitude of the Type II irregularity drift velocity varied over a wide range. The
results of a typical experiment are shown in Figure 10 (spectral data, solid line;

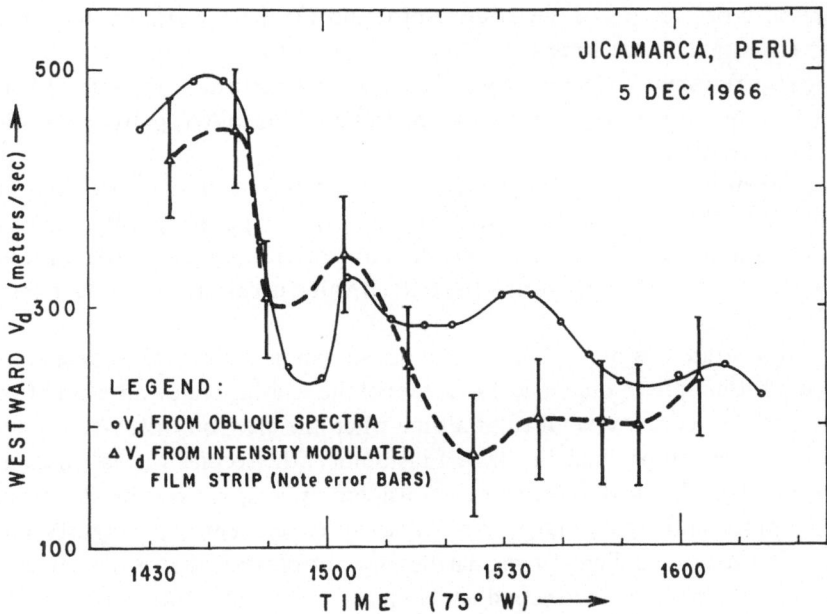

Fig. 10.   Comparison between Type II irregularity drift velocity determined by spectral analysis
(solid curve) and the drift velocity of the irregularity 'patches' (dashed curve).

film-strip data, dashed line). The error bars on the film-strip data indicate the possible
range of error in measuring the line slope.

Figure 10 shows that, even though the drift velocities of the patches and the Type II
irregularities are similar, they are not exactly the same (for example, the curves differ
by about a factor of two at 1530 hours). This is an interesting feature, since, from
scattering theory, both 50 MHz signals must be returned from the same-sized irregularities. This point is perhaps well illustrated here, since the two types of data are
obtained simultaneously from the same receiver output.

That two different 'drift velocities' can be obtained from the same radar signal is
due to the manner in which the signal is treated. The spectral results are a composite
from the total sampled region, with each irregularity contributing equally to the total

power received. The film strips, on the other hand (owing to the non-linear response of the film and the manner in which they were taken), tend to discriminate against the irregularities not contained within a large volume that retains its identity for at least a few seconds. In other words, the spectral results show the 'unweighted' average of the motions of the small irregularities, whereas the film-strip data represent the 'weighted' average of those irregularities whose motions are at least somewhat spatially coherent.

F. IRREGULARITY TYPES AND THE ELECTROJET BIFURCATION

The type of irregularities present in a bifurcated electrojet may be determined from simultaneous data on (1) the vertical structure of the echoing region (to establish the presence or absence of the bifurcation), and (2) the spectral content of oblique echoes (to establish the irregularity types). The vertical structure is obtained as in Balsley (1965), using the Observatory's large, narrow vertical beam, incoherent scatter antenna along with fairly narrow (15 microsec) transmitter pulses and wideband (60 kHz) detectors.

Data taken in this manner show that the Type II irregularities are the only ones present when the electrojet is bifurcated, the Type I appearing only when the two layers are well merged. In other words, the bifurcation occurs only when the electron drift velocity is considerably below the critical value necessary to produce the Type I irregularities.

To demonstrate this point, two spectra are shown in Figure 11. Almost concurrent A-scope photographs of the vertical structure of the echoing region appear in Figure 12. The first of the spectra was taken at a time when the electron drift velocity was high but not sufficient to produce the Type I instability; the second was taken later in the afternoon when the drift velocity was considerably less, as seen by the decrease in mean Doppler shift. Note that in the first set of concurrent data, Figures 11(a) and 12(a), even though the Type I irregularities are not present, the vertical extent of the region has not split. In the second set, Figures 11(b) and 12(b), however, the split region is readily apparent.

G. 'NONTYPICAL' SPECTRA

Occasionally a well-defined peak is observed in the spectra of Type II echoes, which consistently lies near the lower limit of observed Doppler shifts. This phenomenon is shown in Figure 13. The peak seems to be superimposed upon the normally wider spread of Doppler shifts, and tends to occur when the drift velocity is quite low and the echoes are correspondingly weak. It is not a transient phenomenon (e.g. meteor echoes) since the peak can remain for some tens of minutes on typical records. The peaks are seen on approximately 20% of the spectra taken when the drift velocity is low.

H. SYNOPSIS OF THE EXPERIMENT RESULTS

Perhaps the best way of obtaining an overall picture of the electrojet irregularity

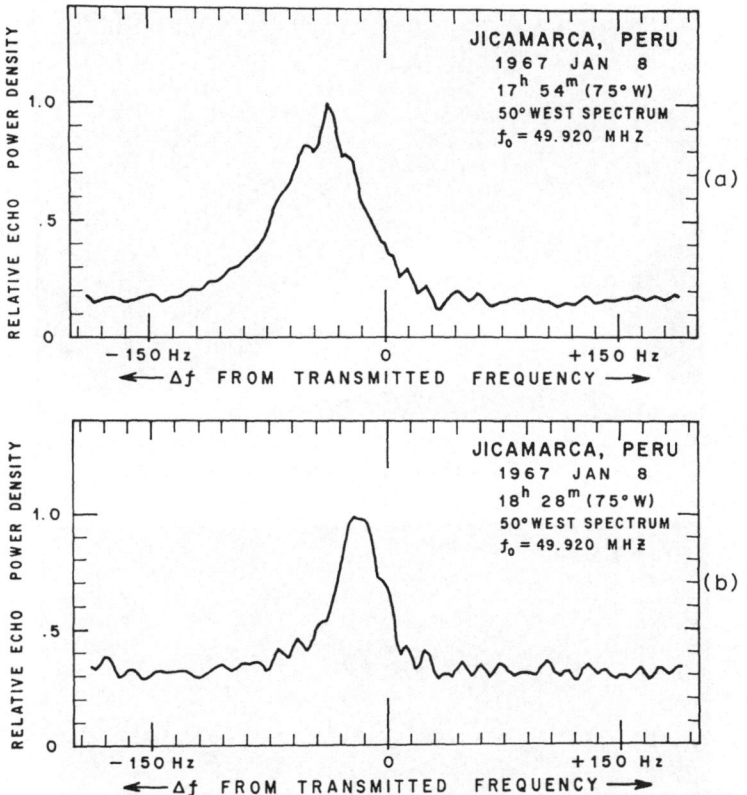

Fig. 11.   Spectra taken with the antenna directed toward the West, in conjunction with Figure 12.
(a) 1754 hours with $\theta = 50$ degrees, (b) 1828 hours with $\theta = 50$ degrees.

characteristics which have been discussed here, is to describe the typical diurnal behavior of the echoing region.

Consider first the period beginning just after local sunrise when the electrojet current is weak and the East–West electric fields are not yet fully developed. Analysis of the weak VHF echoes returned from electrojet irregularities during this period indicates that there are two distinct echoing regions separated in height by about 8 km and essentially equispaced to either side of the conductivity profile (concurrent radar and rocket-borne magnetometer observations have established the height correspondence of the echoing region and the current flow). The mean motion of the irregularities (Type II) present during this period is in the same direction as, and with the approximate velocity of, the westward-drifting electrons. (Superposed upon this mean drift, the irregularities have a random, fairly isotropic motion in the vertical/East–West plane, whose magnitude is comparable to the drift velocity.)

As the electron drift velocity increases, echoes from the two regions become stronger and the regions gradually grow toward the middle until they merge into a single,

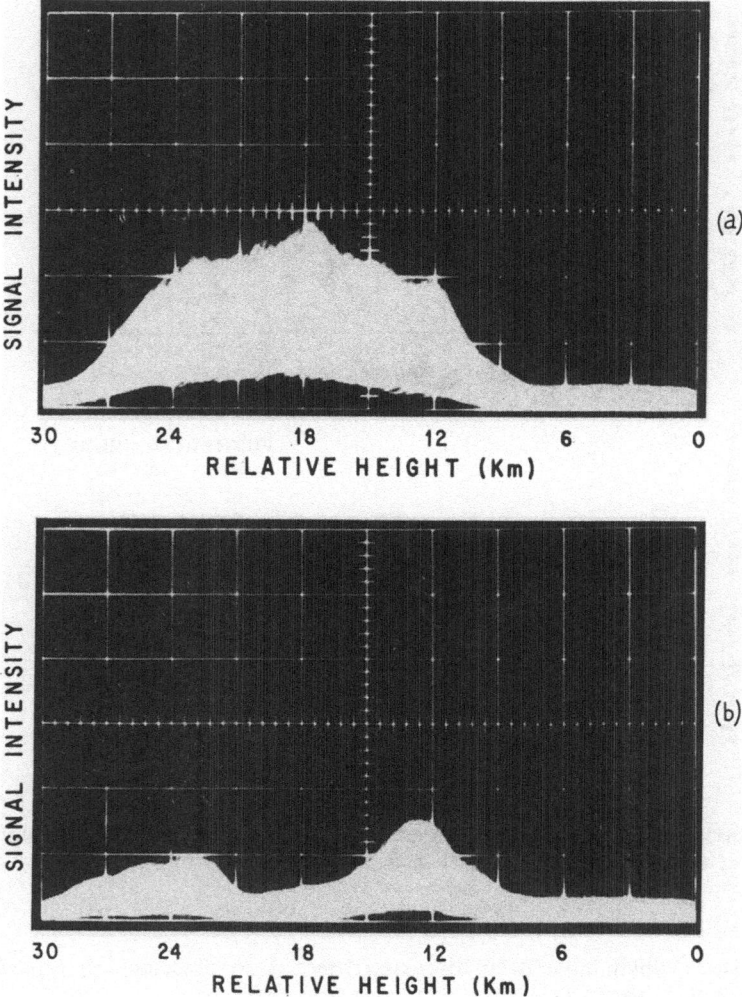

Fig. 12.   Vertical structure of the echoing region taken simultaneously with the spectra in Figure 11
showing that a bifurcated region contains only Type II irregularities.

wide region. Somewhat later, the electron drift velocity becomes comparable to the
ion-acoustic velocity and the Type I echoes appear on the spectra. Typically, this
occurs between 0900 and 1100 hours local time. Further increases in the drift velocity
result in a greater fraction of the total power being returned from these irregularities.
The power returned from the Type II irregularities also increases with increasing drift
velocity, although their echo power (at very oblique antenna angles) becomes quite
small relative to the Type I echo power.

At least during periods when the echoing region is not bifurcated, the presence of
large (typical dimensions of about 1 km) patches of the small (3 m) irregularities seen
by the VHF radar are observed within the echoing region. These patches drift hori-

zontally at velocities comparable to the electron drift velocity and have lifetimes of about 20 sec.

Essentially the reverse of the above sequence occurs during the afternoon period as the current decreases. Decreasing electron drift velocity yields a decreased echo power until the point at which the Type I echo no longer predominates on oblique spectra. The two-stream echo drops out when the electron drift velocity no longer

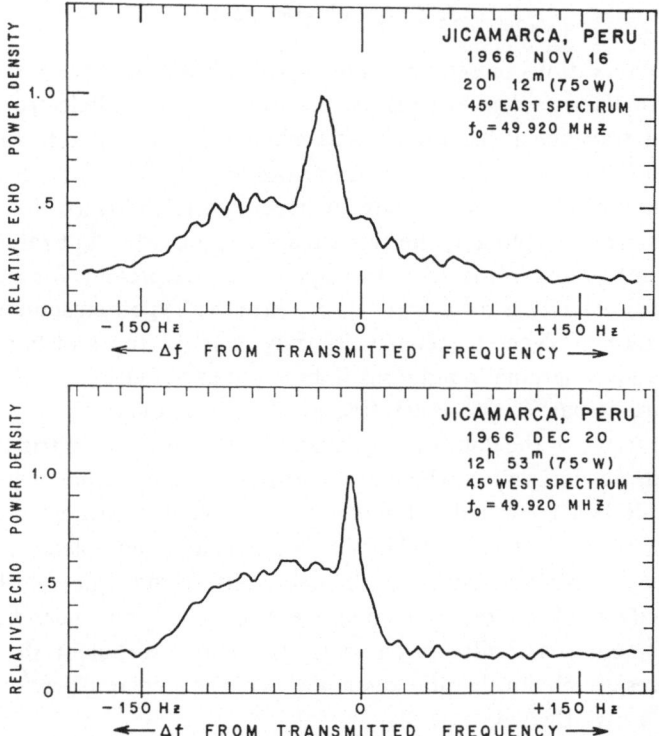

Fig. 13. Nontypical spectra showing the sharp peak which occasionally appears near the center of the spectrum. Both a nighttime and a daytime spectrum are shown, indicating that the phenomenon is independent of electron flow.

exceeds the ion-acoustic velocity. Somewhat later in the period of decreasing drift velocity, the electrojet region exhibits the split character described above, and echoes from the two regions gradually decrease in intensity until they disappear entirely. The electron drift at this time is essentially zero.

Shortly after the electrons have ceased their westward motion they begin to move again. Now, however, the motion is toward the East. The drift velocity increases and the echo power from the Type II irregularities increases accordingly. The eastward motion continues throughout the night. All the characteristics of the daytime echo spectra are observed, as mirror images, in their nighttime counterparts. The nighttime drift velocity often becomes sufficient to generate the Type I instability, and the position of the Type I echo on nighttime spectra shows undeniably that the electrons are now flowing toward the East.

During both daytime and nighttime periods, when the drift velocity is low, the presence of a well-defined second peak is occasionally noted on the spectra. The average frequency shift of this peak is consistently lower than the average shift of the overall spectrum.

3. Concluding Remarks

The series of observations presented in this paper indicates that there are at least two distinct types of electron density irregularities in the equatorial electrojet: (1) planar wave structures, traveling at the ion-acoustic velocity and generated by the two-stream instability, and (2) a different set of irregularities which drift with the electron flow. The general nature of the non-two-stream irregularities indicates that they do not arise from nonlinear coupling of the two-stream irregularities, but rather as a result of some other mechanisms (the Type II irregularities are present, for example, when the electron drift velocity is insufficient to produce the Type I structures).

The appearance of Type II irregularities only near the top and bottom edges of the electrojet under marginal conditions indicates that the electron-velocity gradient plays an important part in the generating process. This quantity may be expressed equivalently in terms of the conductivity profile or the vertical electric field strength. No other parameter has values which are consistently comparable near the edges of the region, while being markedly different near the center. Moreover, this characteristic, plus the other Type II irregularity characteristics (a complete range of irregularity sizes – up to sizes comparable to the half-width of the region – which move at the 'fluid' velocity and have 'random' components normal to the mean flow), suggests that the phenomenon is similar in character to neutral turbulent shear flow. The underlying forces are obviously quite dissimilar. This particular aspect of the problem is being actively investigated.

The West to East motion of the non-two-stream irregularities (and, thereby, the electrons), during nighttime periods, has demonstrated the existence of the nighttime electrojet system. Although the electron velocities are roughly comparable to those of the daytime periods, the reduced electron density precludes the observation of the phenomenon on standard magnetometer recordings. Continuous observations of this nighttime current system may be used to determine particular features of the system itself, and to afford a comparison with the extant S_q system theories.

Since these techniques may be used to obtain the mean drift velocity of the electrons, and since the measurements can be made almost continuously, it is possible to determine the diurnal variation of this drift velocity within the limits

$$0 < V_d (\cong V_e) < V_{ia}. \tag{3}$$

Studies made, for example, during magnetically quiet vs. active periods will yield information on the variation of drift velocity as a function of magnetic activity.

The restriction stated in (3) is not too confining, since the drift velocities – at least during the present period of observation – exceed the ion-acoustic velocity for only a few hours around normal midday periods, and only occasionally at night.

Measurements of the electron drift velocity in this manner may also be used to infer the value of the primary electric field at these heights. Sugiura and Cain (1966) show that the relation between the electron drift velocity V_e and the East–West electric field E_y is given by the approximate expression

$$V_e \simeq - 3.8 \times 10^4 (\sigma_2/\sigma_1) E_y \text{ m/sec}, \tag{4}$$

where σ_1 and σ_2 are the Pedersen and Hall conductivities, respectively. Since the conductivity ratio is about 30 in the region where the electron flow is maximum (Sugiura and Cain, 1966) we may express (4) as

$$E_y \simeq - (8.8 \times 10^{-7}) V_e \text{ volts/m}. \tag{5}$$

The present observations were obtained in the equatorial region where the scattering geometry is relatively simple. It would appear, however, that both types of irregularities are not necessarily confined to equatorial regions, but could exist in similar regions where the electron flow is at least of comparable magnitude. A similar situation exists, for example, within the E regions of the auroral zones, where the electron flows normal to magnetic field lines are occasionally quite intense. It may be that the radar echoes obtained from irregularities in the auroral regions arise from the two types of irregularities which also occur in the equatorial region.

Acknowledgements

It is a pleasure to acknowledge the many interesting and helpful discussions with Drs. D. T. Farley, K. L. Bowles, R. Woodman, T. E. VanZandt, W. B. Hanson, V. L. Peterson, and Professor W. L. Flock. Appreciation is also gratefully extended to the rest of the staff at the Jicamarca Radar Observatory for help during various phases of this work. This work was partially sponsored by NASA under fund transfer R-06-021-008.

References

Balsley, B. B.: 1965, 'Some Additional Features of Radar Returns from the Equatorial Electrojet', *J. Geophys. Res.* **70**, 3175–3182.

Balsley, B. B.: 1966, 'Evidence of the Nighttime Current Reversal in the Equatorial Electrojet', *Ann. Geophys.* **22**, 460–462.

Bowles, K. L., Balsley, B. B., and Cohen, R.: 1963, 'Field-Aligned E-Region Irregularities identified with Acoustic Plasma Waves', *J. Geophys. Res.* **68**, 2488–2501.

Bowles, K. L. and Cohen, R.: 1962, 'A Study of Radio Wave Scattering from Sporadic-E near the Magnetic Equator', in *Ionospheric Sporadic-E* (ed. by E. K. Smith and S. Matsushita), Pergamon, London, pp. 51–77.

Cohen, R. and Bowles, K. L.: 1967, 'Secondary Irregularities in the Equatorial Electrojet', *J. Geophys. Res.* **72**, 885–894.

Dougherty, J. P. and Farley, D. T.: 1967, 'Ionospheric E-Region Irregularities produced by Non-Linear Coupling of Unstable Plasma Waves', *J. Geophys. Res.* **72**, 895–901.
Farley, D. T.: 1963, 'A Plasma Instability resulting in Field-Aligned Irregularities in the Ionosphere', *J. Geophys. Res.* **68**, 6083–6097.
Maeda, K. and Kato, S.: 1966, 'Electrodynamics of the Ionosphere', *Space Sci. Rev.* **5**, 57–79.
Sugiura, M. and Cain, J. C.: 1966, 'A Model Equatorial Electrojet', *J. Geophys. Res.* **71**, 1869–1878.

LOW FREQUENCY WAVES AND IRREGULARITIES IN THE AURORAL IONOSPHERE AS DETERMINED BY RADAR MEASUREMENTS

WALTER G. CHESNUT

Stanford Research Institute, Menlo Park, Calif., U.S.A.

1. Background

For many years it had been apparent that high-frequency communications were disturbed by auroral or magnetic activity. It was also known that signals could sometimes be propagated over long distances by way of reflections off aurora. In the early 1950's experiments showed clearly that radar echoes could be obtained from

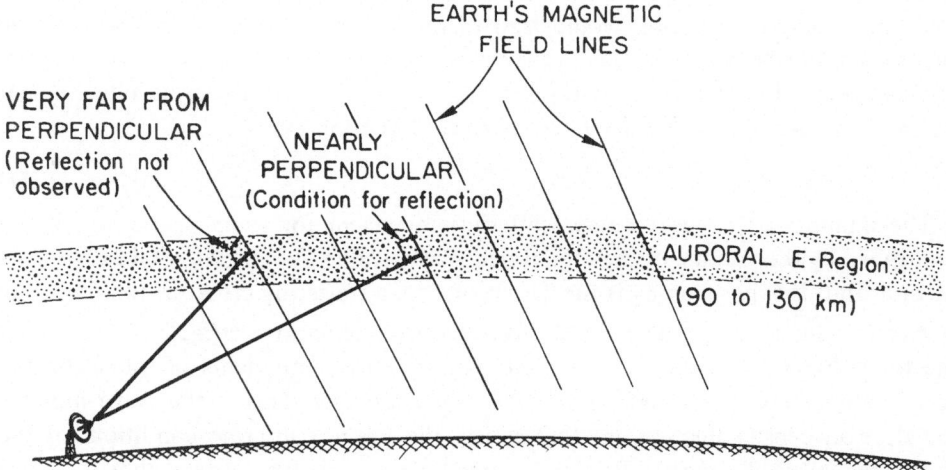

Fig. 1. Illustration of the magnetic aspect requirement for reception of radar echoes from the auroral ionosphere.

the auroral ionosphere during magnetic disturbances. Some of these early data also suggested to Dr. Sidney Chapman (1952) that radar echoes could only be obtained when the line of sight from the radar to the region where the scattering was thought to occur made nearly a perpendicular angle with the earth's magnetic field in the scattering region. This geometry is illustrated in Figure 1. Later work has substantiated Dr. Chapman's concept and suggests that this magnetic aspect condition – or perpendicularity condition – must be satisfied within just a very few degrees or no echoes are obtained.

Physical intuition and logic indicated that the radar scattering by aurora was due to the presence of free electrons in the ionosphere. Initially the reflections were assumed

D'Angelo (ed.), Low-Frequency Waves and Irregularities in the Ionosphere. All rights reserved

to be of an overdense nature – in a manner similar to reflections observed by vertical-incidence ionospheric sounding equipment. When radar echoes from aurora were obtained at frequencies above 100 MHz, it became apparent that this explanation of overdense plasma for the auroral echoes was not tenable. A plasma that is overdense at 100 MHz contains an electron density that exceeds 10^8 per cm^3.

In 1956 Professor Henry Booker (1956) applied Born approximation scattering theory to the auroral ionosphere in an attempt to explain the scattering from a plasma that was far from overdense at the radar frequency employed. Booker's picture of the ionosphere was one in which some mechanism, at that time unknown to him, produced fluctuations in the electron density in the ionosphere. These fluctuations in electron density would vary more gradually along the earth's magnetic field lines than they would across the earth's magnetic field lines. The description used by the auroral community to relate to Professor Booker's concept was one of cigar-shaped blobs of high electron density aligned along the earth's magnetic field – though this picture is not viewed as being close to the true situation as we know it. This anisotropy in electron density distribution would lead to radar scattering that required that the radar beam line be perpendicular to the earth's magnetic field in the scattering region, as was observed experimentally. Booker's expression for the scattering intensity per unit volume per solid angle would be given by Equation (1):

$$\sigma_v = \sigma_T \overline{\Delta n_e^2} \Phi(k_r, k_\theta, k_z). \tag{1}$$

The quantity σ_v is the volume scattering coefficient for radar scattering. When multiplied by the volume of the region that scatters, one obtains the total radar cross section. In this expression σ_T is the Thomson radar scattering cross section for a free electron*; $\overline{\Delta n_e^2}$ is the variance in electron density fluctuation; $\Phi(k_r, k_\theta, k_z)$ is called the three-dimensional spectral function. The various components of the scattering wave number are given here in cylindrical coordinates. This is the wave number for electron density fluctuations that satisfy the Bragg scattering condition for the radar frequency of concern. Booker assumed azimuthal symmetry so that k_θ would not be used.

Booker's real contribution to the whole art of auroral radar scattering was the introduction of a scattering concept that would relate the nature of observed radar echoes to the scattering medium. This is very important in that experimentalists could begin to talk about the nature of the medium rather than just about voltages, etc., in their radar receivers.

At the time Booker introduced Born approximation scattering to radar physicists, he did not have a good explanation for the origin of the fluctuations, which he found were necessary to explain the auroral radar measurements. It is the origin of these fluctuations that determines the precise form that one should use for the 3-dimensional

---

* Radar cross-section is defined as the differential scattering cross section for scattering at 180° (backscatter). Unlike cross-sections used in other disciplines, the differential scattering cross-section is defined as that *per 4π steradian* rather than per steradian. Thus, the value of σ_T that one uses in Equation (1) is 4π times the value for backscatter obtained from books on atomic physics.

spectral function Φ. Further radar auroral work was performed by many organizations in the hope that a study of the relationship between radar echoes and other features of the aurora borealis would reveal the form of the function Φ and the mechanism by which fluctuations in electron density are produced.

During the period 1960 to 1963, radar experiments performed at Jicamarca on the equator by Bowles *et al.* (1963) showed that the equatorial electrojet, which flows at an altitude of 100 to 120 km, gave rise to radar scattering that in many ways resembled auroral radar scattering. In fact, the most telling feature was that the magnetic aspect condition, which is so pronounced in auroral scattering measurements, was also observed in the equatorial measurements. Careful experiments by this group at the equator showed that a reasonable explanation for the scattering that they observed was one that hypothesized longitudinal waves, or density waves, in the several-meter wavelength region traveling in the equatorial ionosphere. They found by Doppler-shift measurements that these waves traveled with a velocity that was nearly sonic, which means about 300 to 400 m/sec. This velocity was remarkably similar to that measured by Doppler shift in auroral scattering.

The experimental observation at the equator was later provided with a theoretical explanation by Farley (1963), who showed that, under the proper conditions, when currents flow in the equatorial ionosphere one might expect that longitudinal plasma waves could grow. Farley showed that these waves would grow in directions such that the component of the streaming electron velocity in the wave direction would exceed a critical threshold value. If the component of streaming electron velocity was less than this threshold, no waves would grow. Most importantly, Farley's theory showed that waves would only grow nearly exactly perpendicular to the earth's magnetic field. This very important feature is essential in understanding equatorial electrojet scattering. This work by the Jicamarca group now provided a reasonable explanation for the density fluctuations that Booker had hypothesized several years before.*

One of the difficult features of auroral radar work is that, unlike the equatorial electrojet, the auroral ionosphere presents a very complex current flow structure. Thus, if one wishes to study longitudinal plasma waves by radar, the auroral ionosphere presents many complexities that are nonexistent at the equator. Continued experiments in the auroral ionosphere have convinced us that the explanation obtained from the work at the equator does apply to auroral radar scattering. We do not yet know whether other phenomena, other types of instabilities, or other scattering mechanisms play a role in auroral scattering.

Continued work has been oriented toward finding relationships between the auroral radar scattering and other auroral features such as visible luminosity, magnetic disturbance indices, particle precipitation, and many other auroral zone phenomena. Our own experiments that will be presented here have been directed primarily at determining the function $\Phi(k_r, k_\theta, k_z)$.

* More recent measurements presented by Dr. Balsley in this volume now show that the situation in the equatorial electrojet is more complicated than the picture presented here.

2. Introduction

Experiments over the years have used radar reflections from the auroral ionosphere to keep track of the location of auroral activity. These researches have generally assumed that a close relation exists between radar aurora and optical aurora. Indeed there exists some strong evidence that for low frequency radars with frequencies as high as 20 to possibly 50 MHz, there may be a fair degree of correlation between radar and optical aurora (Bates *et al.*, 1966). In the particular work cited, the correlation seems close but not exact. The data may also suggest that scattering mechanisms other than that discussed in Section 1 are important at these low frequencies, which means that the observed correlations are not applicable to the higher frequencies that are pertinent in later portions of this report. Experiments by Bowles (1955) at 106 MHz showed a variety of degrees of correlation between radar and optical aurora from very close correspondence to none at all. We shall present later some of our results on this question.

Other experiments have been performed to relate the presence and intensity of auroral echoes to other geophysical quantities such as local and planetary k indices. Most notable of the studies are the very important experiments and analyses performed by Unwin and Know (1968) at 55 MHz – surprisingly enough in the Southern hemisphere. For the reader interested in pursuing this aspect of radar aurora, Mr. Unwin's papers also present an excellent review and a rather extensive reference list.

The purpose of our discussion here is not to recount and classify the entire morphology of radar aurora – but rather to discuss the nature of the scattering medium and the aurora's irregularity structure as we think we now understand it. That radar and optical aurora are both manifestations of magnetic disturbances is now clear. That optical emissions are associated with particle precipitation seems to be generally accepted. That auroral radar reflections at frequencies above, probably, 50 MHz and, certainly, above 100 MHz result from Bragg scattering of the radar waves from fluctuations in electron density in the ionosphere primarily between 100 and 130 km altitude also seems to be universally accepted. Whether the region of particle precipitation and the area of small-size (less than several meters) inhomogeneities in electron density responsible for radar scattering coincide or are displaced has not been resolved. We intend to present below our latest findings that bear on this problem. We also present our latest results concerning the structure of the function $\Phi(k_r, k_\theta, k_z)$ of Equation (1).

3. Experiment Geometry and Instrumentation

Figure 2 presents a schematic map of the Northern regions of the earth. Shown on this map is the auroral zone after Vestine (1944). Auroral radar scattering experiments have been performed at UHF radar frequencies by the Radio Physics Laboratory of the Stanford Research Institute at the three locations shown. The approximate periods of data collection are also indicated. The magnetic aspect – or perpendicularity – condition restricts the region from which radar echoes can be obtained from each site to the areas that are shaded. The data to be presented below were obtained by our auroral radar observatory at Homer, Alaska. The region from which radar echoes can be obtained from the Homer observatory coincides quite well with the region

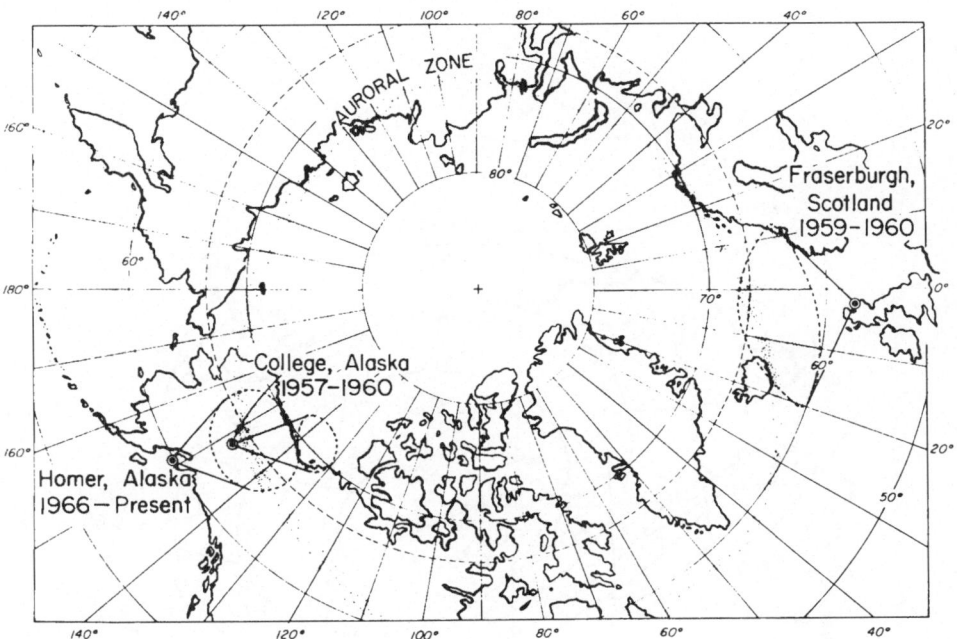

Fig. 2.   Map of Northern polar regions indicating locations and regions of echo probing for various auroral radar observatories operated by Stanford Research Institute.

of statistically frequent auroral activity. As is also evident, the region from which echoes are expected from the Homer observatory coincides very nicely with the region of Alaska that is well instrumented for other auroral experiments by the Geophysical Institute of the University of Alaska.

The most important thing to keep in mind about auroral radar scattering is the following: the Born approximation theory developed by Booker that we use to interpret our radar measurements shows that the radar signal essentially Fourier-analyzes the scattering medium along the radar beam line. If there are electron density fluctuations whose spatial wavelength is equal to half the radar wavelength, then backscatter takes place and the radar receiver indicates a target. In the case of the traveling plasma waves in the ionosphere that we believe are responsible for auroral scattering, the radar observes only those plasma waves that travel exactly towards – or away from – the radar. These waves must also have a wavelength that is equal to one-half the radar wavelength in order that energy be scattered back to the radar. This is, in essence, the Bragg condition. Thus, a radar probing of the auroral ionosphere is in a sense a plasma wave detection system. The function $\Phi(k_r, k_\theta, k_z)$ of Equation (1) then really tells us how intense the plasma waves are that have wavelengths and a direction of travel indicated by k_r, k_θ, and k_z. That is, it is the three-dimensional spectral distribution function for plasma waves. This picture that we have presented may not be completely true; nevertheless, it is a convenience for organization purposes.

The work at Homer, Alaska, has concentrated primarily on the determination of

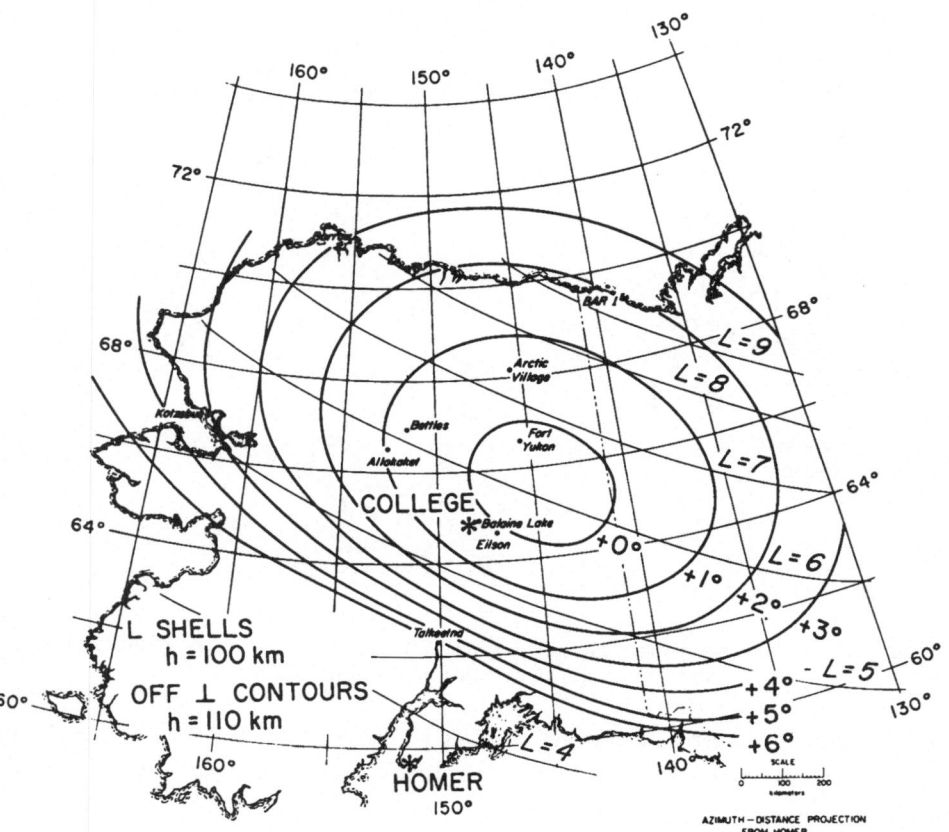

Fig. 3.   Map of Alaska showing contours of quiescent magnetic aspect angle at an altitude of 110 km as observed from the Homer radar site.

the radar frequency dependence of radar backscatter echoes. It was hoped that this determination would aid in developing a further understanding of the nature of the plasma wave structure at altitudes on the order of 110 km where echoes are by far predominant. We have also concentrated on determining the magnetic aspect angle dependence of our auroral reflections. The magnetic aspect dependence is how the radar scattering intensity decreases as the magnetic aspect angle deviates from exact perpendicular. These two quantities, the frequency dependence and the magnetic aspect dependence, essentially define the function $\Phi(\mathbf{k})$.

Some of the Homer radar data have been correlated with all-sky camera photographs of aurora. We have not been concerned with the general morphology of radar backscatter with the Homer radar.

Figure 3 presents a map of Alaska. Homer is located near the bottom center of the figure. The figure also presents the contours over Alaska along which the radar beam makes an angle of 0°, 1°, 2°, etc. with the perpendicular to the earth's magnetic field at a height of 110 km. This is the height at which most scattering takes place. The 0° contour is the region where one would most expect auroral echoes. Most echoes will be obtained from within the 2° to 3° contour. The magnetic field assumed is the undisturbed, quiescent magnetic field. It is well known that the earth's magnetic field can be distorted by several degrees at an altitude of 100 km during magnetic disturbances. Therefore, during magnetic disturbances these contours must necessarily change.

Figure 4 is a photograph of the Homer radar. The Homer radar includes six frequen-

Fig. 4.   Photograph of the Homer radar showing 20-m parabolic antenna and transmitting and receiving buildings.

cies from 50 MHz to 3000 MHz. Table I presents a brief summary of the radar characteristics. The 6 frequencies will sense the presence of longitudinal plasma waves of wavelengths 3, 1.07, 0.38, 0.18, 0.12, and 0.05 m. The radar pulse length that is normally used is 300 μs. This 300 μs pulse length means that at any instant energy will be received that could have come from any place along the radar beam line over a distance of about 45 km. This range depth is much greater than one would like, but it is necessary in order to obtain the detection sensitivity required to make meaningful measurements of the aurora. The long pulse means that fine spatial details or structure in the auroral scatterer will be blurred or not seen at all.

The Homer radar antenna is a fully steerable 20-m-diameter paraboloid. All

TABLE I

Homer auroral radar characteristics

Frequency (MHz)	50	139	398	850	1210	3000
Peak power (kW)	30	50	40	40	40	200
Pulse length			300 μs all frequencies			
PRF			75 Hz all frequencies			
Antenna			Steerable 20-m parabola on all frequencies			
Ant. beam width	24°	9°	3°	1.5°	1°	0.4°
Receiver bandwidth	10 kHz			26 kHz		
Receiver dynamic range			> 60 dB			

frequencies are fed through this dish. Therefore, each frequency has a different antenna beam width. Accordingly, there are times when we are not certain that all frequencies are observing the same auroral feature. This is very undesirable, but necessary from a practical point of view.

4. Cross-Correlation between Optical Photographs and Radar Returns

Since we believe that auroral radars are detecting longitudinal plasma waves, cross-correlation of the features of auroral echoes with other auroral phenomena then relates ionospheric currents and plasma waves to the same phenomena. Jim Hodges of our laboratory has made an analysis of all-sky photographs of aurora taken by the University of Alaska on two separate nights during which we received radar echoes. The first night he analyzed was one of very active aurora and intense radar echoes. It was also a night during which the aurora wandered all over the sky. On this night no detailed correspondence between visible and radar aurora was found, other than the fact that they were both very active and pretty much covered the sky. Both photographic and radar data were acquired over a period of many hours. The camera used sensed the aurora once a minute. One complete all-sky survey by the radar took 7 min. The auroral structures changed completely during both these short data-sampling intervals.

On the night of 16 March 1967, the auroral activity was considerably weaker. Auroral forms were relatively stable. It would not be suitable to present all the comparisons prepared. In summary, though, more often than not there were visual auroras not accompanied by any radar aurora. However, Figure 5 shows a relationship obtained on one radar scan that has been seen fairly often. The visual feature at this instant is a single, long arc north of Fort Yukon, denoted here by the squiggly line.

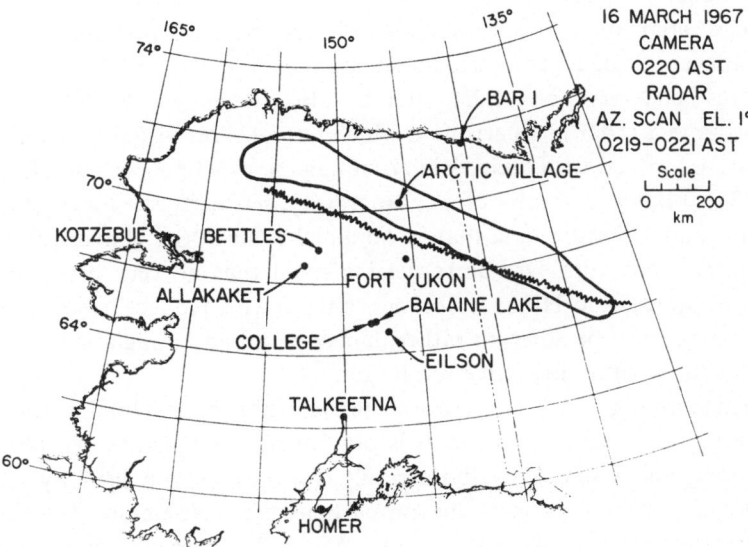

Fig. 5.   Radar-optical correlation map – 16 March 1967, 0220 Alaska Standard Time. A visual auroral arc was located by triangulation between two cameras and is illustrated by wavy line. Auroral radar echoes originated from within indicated contour.

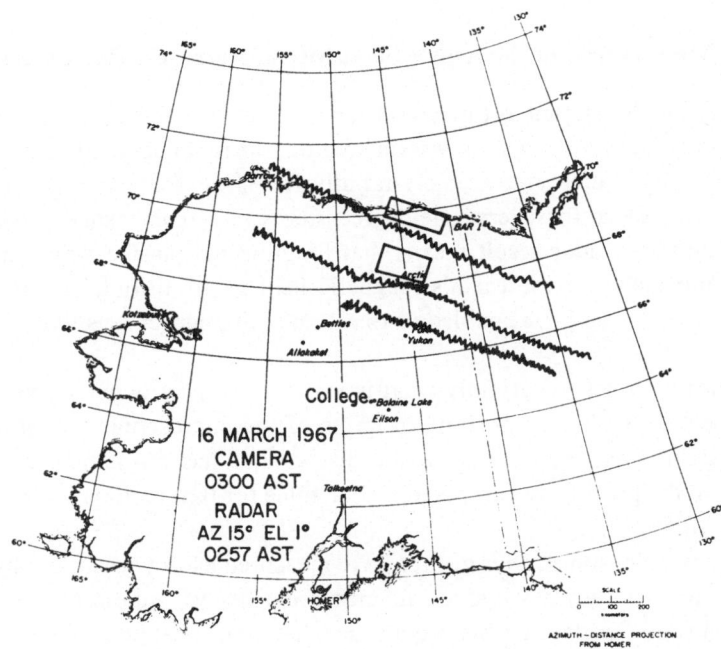

Fig. 6.   Radar-optical correlation map – 16 March 1967 0300 Alaska Standard Time. Three visual arcs were located by photographic triangulation. Radar beam was held at fixed position; radar echoes were obtained from inside the two rectangular contours shown. Radar beam width and echo range depth determine contours.

The lower border of this arc was located by triangulation between at least two cameras. The radar auroral scattering arose inside the contour indicated. We note that there appears to be a geometric similarity, but these two regions do not exactly coincide. This behavior was revealed often by Mr. Hodges' study.

Figure 6 shows a radar comparison in which two of three visual arcs seem to have radar echoes associated with their vicinity, but again, exact collocation did not exist. Dr. H. F. (Skip) Bates, currently of our laboratory, has performed a careful analysis of the optical and HF radar data that he obtained while at the University of Alaska. In private discussions with Dr. Bates, he indicated that he very often finds near-collocation of optical and HF radar echoes, but, as with Hodges' results, they more often than not seem to be adjacent rather than superposed, though he does have one case of collocation within his range resolution.

It had been tempting in past years to assume that where optical luminosity occurred, radar scattering also ought to occur. It is our belief now that the two phenomena, radar scattering and visual glow, both result from a common primary effect, i.e., some effect causes precipitation of auroral particles that cause glow. The same effect also gives rise to electric fields in the auroral E-region. These electric fields need not be most intense or necessarily collocated with the precipitating particles, but we believe, after Farley (1963), that they are responsible for the production of longitudinal plasma waves that are the irregularities in electron density from which our radar signals scatter.

5. Measurement of the Aspect Sensitivity of Auroral Radar Echoes

Experiments have shown that as the radar beam line deviates from exact perpendicular to the earth's magnetic field in the scattering region, the radar echo intensity decreases. The echo intensity decreases extremely rapidly for UHF frequencies – by about a factor of 10 in power for every one-degree step away from exact perpendicular. We assume that this radar result means that longitudinal plasma waves travel only nearly perpendicularly to the earth's magnetic field at an altitude of 110 km. Our measurements at Homer have enabled us to estimate this aspect sensitivity from 50 to 1210 MHz.

It is important to define carefully what is meant by aspect dependence. When we speak of aspect dependence, we think in terms of the microscopic conditions in the scattering region. That is, if the magnetic disturbance has reoriented the magnetic field lines, then the previously computed aspect angle for that region is no longer valid within the scatterer.

Over the past years some radar groups have measured aspect sensitivity by plotting the average signal intensity obtained in measurements vs. quiescent aspect angle. We have organized our data in this way to see what would happen. Figure 7 presents our data organized on a statistical basis and plotted against quiescent magnetic field aspect angle. We have plotted vertically the logarithm of the volume scattering coefficient vs. aspect angle as abscissa. We have plotted only the 139 MHz and

850 MHz data, since the 398 and 1210 data are similar. Note that the volume scattering coefficient is large over several degrees near exact perpendicular.

We hypothesize that when aurora is weak the magnetic field is undistorted, so echoes are obtained from regions of zero aspect angle (exact, quiescent perpendicular).

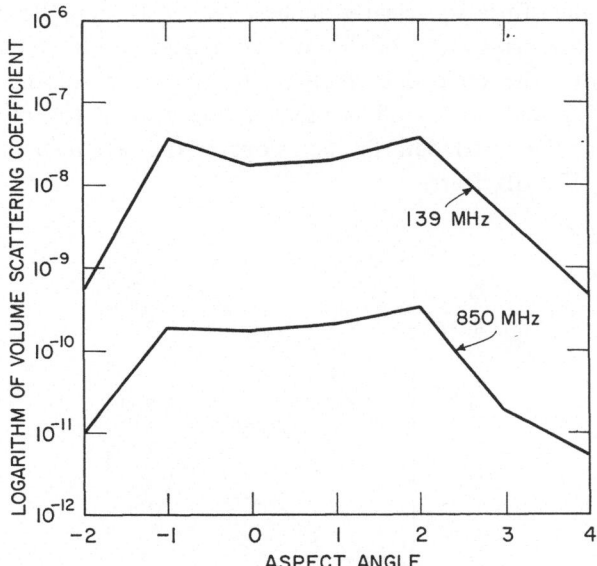

Fig. 7.   Logarithm of volume scattering coefficient vs. quiescent magnetic aspect angle in echoing region. Data summarizes 497 scans of aurora by Homer radar.

As the intensity of the magnetic disturbance increases, the magnetic field becomes distorted, so new regions, formerly somewhat off perpendicular, now become regions of microscopic perpendicularity. Because the magnetic disturbance is stronger, echo amplitudes are stronger. Thus, we believe that the data in Figure 7 represent some sort of convolution of a steep aspect dependence with distortions of the earth's magnetic field in the scattering region during periods of magnetic disturbance.

In order to determine the microscopic aspect angle dependence, another procedure must be used. It is this magnetic aspect dependence that is needed for comparison with predictions of plasma wave behavior in the ionosphere. On the night of 24 November 1967 a very stable radar arc was located by the Homer radar. This arc was well defined by the narrow-beam 850 and 1210 MHz radars. Figure 8 shows the volume scattering coefficient for the four middle radar frequencies vs. azimuth. This plot represents the scattering cross-section as a function of position along the arc. Each position along the arc has a unique magnetic aspect angle. We note that, as well as we can determine, the shapes of these four curves are nearly the same. This similarity implies that the magnetic aspect sensitivity is, at most, a weak function of frequency over this range of frequencies. This conclusion should be accepted as tentative pending more careful, statistical analysis of these data and other data sets as yet unreduced.

Figure 9 presents the logarithm of the 398 MHz signal volume scattering coefficient vs. magnetic aspect angle. In Figure 9(a) we have assumed that there has been a slight uniform distortion of the magnetic field corresponding to a bending of the magnetic field lines at 100 km southward by $\frac{1}{2}°$. We see that excellent consistency is obtained by hypothesizing this magnetic distortion. Other forms of magnetic distortion were attempted; this data form was clearly the best. So, we assume that by this means we have calibrated the orientation of the disturbed field within the scattering region. Figure 9(b) shows the same data plotted vs. the square of the magnetic aspect angle. For the present, we would say that a linear fit of log of cross-section vs. aspect angle gives the better data fit. The slope of the best linear fit to the data in Figure 9(a) is -12.1 dB/degree.

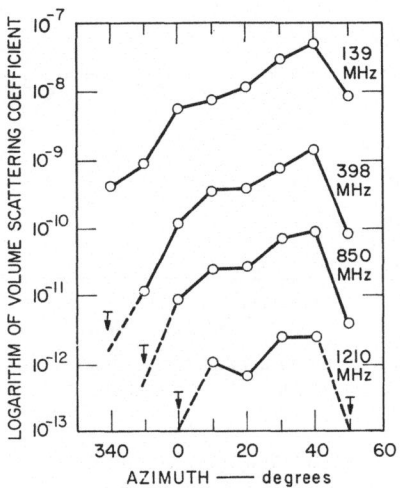

Fig. 8.   Logarithm of volume scattering coefficient vs. azimuth to echoing region, for a stable radar auroral arc, at several radar frequencies.

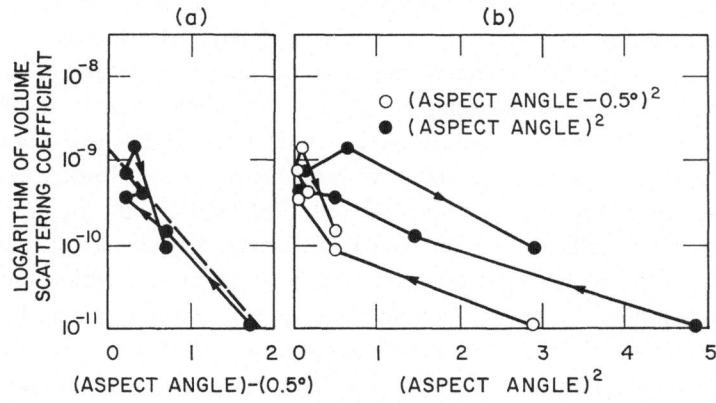

Fig. 9.   Logarithm of volume scattering coefficient vs. various functions of computed aspect angle data obtained at 398 MHz.

The Homer radar does not have sufficient signal strength to measure aspect sensitivity at 3000 MHz. The 3000 MHz echoes have been obtained in regions where the quiescent magnetic aspect angle has been fairly large. The evidence suggests that the aspect sensitivity is the same at 3000 MHz as it is at the four lower frequencies, though this point has not been absolutely proved.

The determination of aspect dependence at 50 MHz is a more difficult chore. The importance of refraction in the E-region ionosphere to 50-MHz aspect measurements has only recently been recognized. The data from this quiescent auroral arc as obtained at 50 MHz reveal only a weak reflection from the region of the arc where all other frequencies were returning maximum signal strength. On the other hand, the 50-MHz signal strength increased in regions of undisturbed aspect angle and peaked in regions of quiescence three to four degrees off exact perpendicular. We have made the assumption that this displacement of position of peak return results from the 50-MHz microscopic aspect dependence being very steep, as it is with other frequencies. Based upon this assumption, we can make a correction for refractivity of the ionosphere and determine the signal strength vs. microscopic aspect angle for 50 MHz. Figure 10 presents these results. These data are smoothed somewhat. The width of the aspect dependence curve is somewhat broader than for the higher frequencies, but is very narrow compared with all prior estimates.

We summarize our Homer radar aspect sensitivity results in the following way:

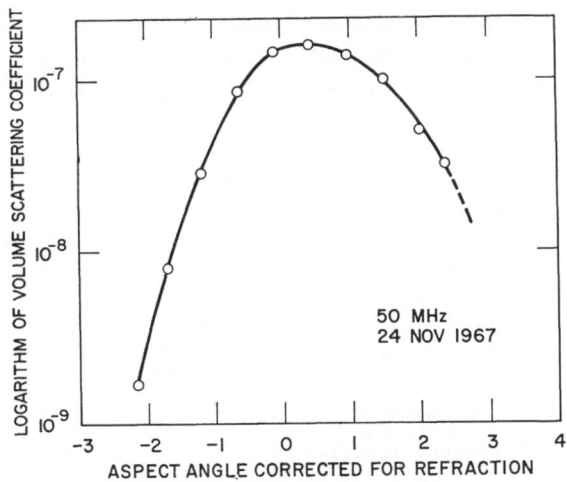

Fig. 10. Logarithm of volume scattering coefficient vs. computed aspect angle. Data obtained at 50 MHz. Aspect angle corrected for refraction in the E-region. Measurement implies electron density of $6.4 \times 10^5/\mathrm{cm}^3$ in scattering region.

it appears that the aspect dependence as measured between 139 MHz and 1210 MHz is only weakly dependent on the frequency employed. Our data obtained at 3000 MHz, which have not yet been displayed, suggest that the dependence at this higher frequency has a similar aspect angle dependence. Provided one accepts the argument that refraction dominates 50-MHz data, then the aspect sensitivity at 50 MHz is not very much weaker than at the higher frequency. These data, if correct, would imply that longitudinal plasma waves at E-region heights with wavelengths from 3 m down to 5 cm all travel in the same direction – nearly perpendicular to the earth's local magnetic field.

There is a serious flaw in the arguments above. It is customary to assume that the radar scattering dependence is cylindrically symmetrical about the earth's magnetic field. Indeed, the data have been analyzed as if this were so. We believe that the magnetic aspect dependence is produced by streaming electrons generating plasma waves traveling in the ionosphere, generally in directions parallel to the current flow, but always almost exactly perpendicular to the magnetic field. This is the plasma wave concept proposed by Farley (1963). If this is so, then our experimental assumption of cylindrical symmetry about the earth's magnetic field, which is used in our aspect sensitivity measurement, is not valid. The result would be that the true aspect dependence is even steeper than that presented in Figure 9.

6. Frequency Dependence of Auroral Echoes

The Homer observatory radar spans nearly $5\frac{1}{2}$ octaves in frequency. Unfortunately, the lowest frequency data at 50 MHz must be specially handled because of the very broad antenna beam pattern and ionospheric refraction. Likewise, signals at 3000 MHz are so weak when they are detectable at all that they must also be handled in a special manner. What this means is that usually we cannot obtain at a given time a complete spectrum of ionospheric plasma waves from 50 MHz to 3000 MHz.

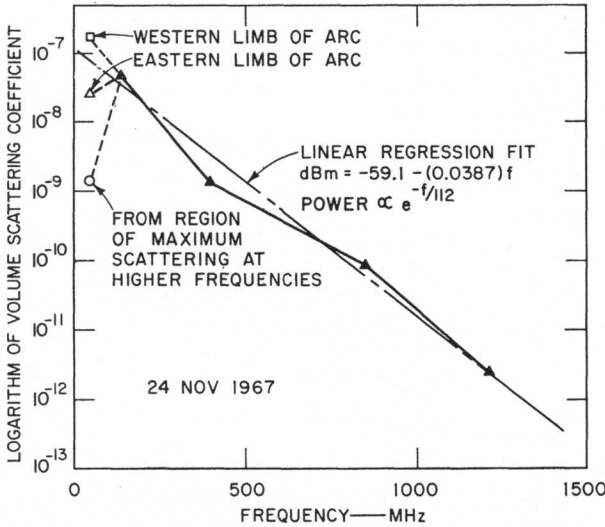

Fig. 11. Logarithm of volume scattering coefficient vs. frequency. Data points plotted are for assumed zero magnetic aspect.

Figure 11 shows the spectrum of plasma waves obtained from the data from which aspect sensitivity was determined. We have plotted the logarithm of the volume scattering coefficient vs. radar frequency in MHz. Thus, in essence, we have plotted logarithm of reflectivity vs. plasma-wave wave-number. It has been found that almost all of the Homer observatory data can be well fit by a straight line on this plotting format. The 50-MHz data are plotted for three regions. The Western and Eastern limbs of the arc mean that that was the scattering cross-section where we assumed that the microscopic perpendicularity condition was met at 50 MHz. The low point is the signal strength at 50 MHz from the region where the other, unrefracted echoes at higher frequencies were maximized. In the region where 50 MHz maximized – in the East and West limbs – no echoes were obtained at the other frequencies at all.

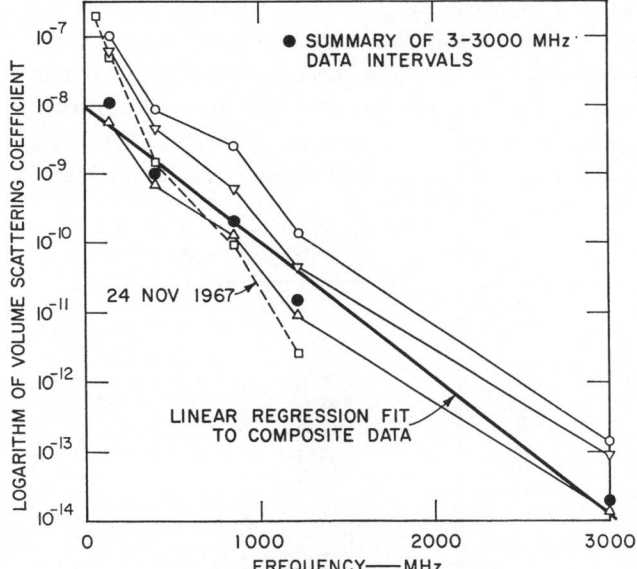

Fig. 12.   Logarithm of volume scattering coefficient vs. frequency for data obtained when 3000 MHz echoes were observable. The spectrum for Figure 11 is superimposed for comparison.

The amplitude of longitudinal plasma waves at wavelengths corresponding to our radar frequencies in backscatter, i.e., with plasma wavelengths equal to one-half the radar wavelength, will be proportional to the square root of the volume scattering coefficient at that frequency. Therefore, the logarithm of the plasma wave amplitude distribution will also plot as a straight line. Thus, the spectral distribution of plasma wave amplitudes will be exponential – like our radar volume scattering coefficient.

Figure 12 presents the received power or auroral radar cross-section vs. frequency for the data when 3000-MHz echoes were received. The three sets of open points represent three separate data collection periods – two of which were on the same evening. The filled circles are the average of all data obtained that contained 3000-MHz echoes. For comparison purposes, the data from the 50-MHz experiment have also been plotted. The slope of the data including the 3000-MHz point is more gradual than that of the data fit on which the 50-MHz echoes were analyzed. A compilation of 40 separate determinations of the auroral frequency dependence – using only the middle four radar frequencies – reveals that the straight-line behavior shown here is normal, but that the slope of the line changes from time to time. We know of no obvious reason for the slope change, nor have we had an opportunity to seek a correlation between the slope and some other geophysical quantity.

The analytical form for volume scattering coefficient vs. radar frequency is

$$\log\{\sigma_v(f)\} = -f/f_0 - \log\{\sigma_v(0)\}$$

or

$$\sigma_v(f) = \sigma_v(0)\exp\{-f/f_0\}, \tag{2}$$

where f is frequency in MHz and f_0 is the reciprocal of the slope of the linear fit. For convenience we shall call f_0 the 'scale' frequency.

Figure 13 presents a histogram for the occurrence of f_0. We have found that the slope can vary by as much as a factor of 2. We also find that this variation of slope now permits us to rationalize apparent inconsistencies in data we have obtained over the past 10 years. When using only two frequencies, it was often convenient to use

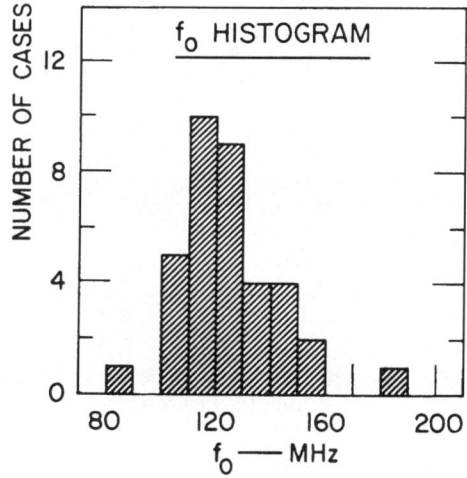

Fig. 13.   Histogram of occurrence of f_0 – scale frequency.

a simple power law fit between the two points. This power law relationship from experiments made at one time often seemed to contradict data obtained at another time. The scale frequency histogram provides us with the realization that – as we measure it – the spectrum of plasma waves does not seem to be time-invariant.

7. Relationship of Our Radar Auroral Results to Longitudinal Plasma Waves

As indicated earlier, the radar backscatter from aurora, in essence, Fourier-analyzes the auroral ionosphere for variations in electron density that are sympathetic with the radar wavelength in backscatter. By Equation (1), the receiver power will be proportional to the square of the ionospheric electron density and to the square of the amplitude of the longitudinal plasma wave. Frequency dependence of the radar backscatter, therefore, tells us the spectral distribution of wavelengths of longitudinal plasma waves traveling toward or away from the radar. The magnetic aspect angle dependence indicates how intense the plasma waves are when they attempt to travel slightly off-perpendicular to the earth's magnetic field.

The interpretation of these amplitude data with regard to streaming electrons in the auroral ionosphere is complicated by the fact that the waves need not travel parallel to the current system. Furthermore, it is now known that the direction of current flow in the auroral ionosphere is a strong function of altitude. Thus the relating of radar scattering intensity to theories of longitudinal plasma wave generation will necessarily be complex.

Perhaps the most interesting feature of the new results presented here is that the spectrum of plasma waves nearly always appears to show an exponential dependence upon plasma-wave wave-number. This suggests that the auroral E-region is an environment in which plasma waves of some wavelength will grow to some amplitude. At this fairly large amplitude the wave will interact with other waves, producing harmonics and cross-modulation products; these have shorter wavelengths and, therefore, lead to scattering at higher radar frequencies. These higher harmonics and cross-modulation products will continue to grow until they begin cross-modulating between themselves and the primordial waves, producing higher and higher frequencies, shorter and shorter wavelength plasma waves.

Our intuition likens this situation to that which exists in turbulence in a fluid medium. There, large eddies, produced by hydrodynamic flow, break up into smaller eddies, which break up into smaller and smaller eddies, etc., and establish a stable spectral form. In normal fluid turbulence the spectral form is not very dependent upon input parameters, provided the turbulence is sufficiently intense. We are suggesting here that the recently analyzed data from the Homer radar measurements are revealing a similar kind of phenomenon for longitudinal plasma waves in the auroral zone.

8. Summary

In this report we have summarized the development of our current understanding of radar scattering from the aurora borealis at radar frequencies above 50 MHz. Past radar experiments performed in the auroral regions and in the equatorial electrojet have indicated that radar signals are reflected from plasma waves travelling with wavefronts that are perpendicular to the earth's magnetic field. These plasma waves probably result from amplification of noise-like background by streaming electrons that flow in the 100 to 130 km altitude region during magnetic disturbances.

Past comparison of the association of radar aurora with optical aurora has shown correlation in location that varies from none whatsoever at some times to very close at other times. Measurements presented here show that during one night of quiescent, stable aurora, a correlation in the shape of the optical aurora and of the radar reflecting region did seem to exist, but the two regions were not collocated. This result suggests that the strongest electric fields at *E*-region heights that produce high velocity electron streams and plasma waves might lie outside visual auroral features. Magnetometers indicate that the main currents flow within the visual features. This is not necessarily a contradiction, since plasma waves are produced, according to Farley, by high-speed electrons accelerated by strong electric fields. Large currents can flow in regions of high conductivity (high electron density) with rather moderate electric fields.

The research results presented here show that the magnetic aspect sensitivity of auroral radar reflections is weakly dependent upon radar frequency from 50 to 3000 MHz. In fact, there may be no frequency dependence. This result suggests that the shorter wavelength plasma waves that reflect higher frequency radar waves are not produced directly by streaming electron amplification. Rather, these waves may be produced by cross-modulation of longer-wavelength waves traveling in the same direction. Thus, their short wavelength progeny will all travel in the same direction as the primordial waves and with approximately the same velocity – as would be required to be consistent with our Doppler measurements (not presented).

The distribution of plasma-wave wavelengths appears to be exponential in wave-number, as revealed by our radar probing. The 'scale frequency' for the exponential form changes from time to time. The reason for the apparent change in scale frequency is not known.

It is hoped that further radar experiments in both the auroral and equatorial regions will be performed. Diverse frequency measurements at the equator could reveal the plasma-wave spectral distribution in a way that might shed light on nonlinear plasma processes. Careful Doppler measurements at higher frequencies in the auroral region may reveal the extent to which instability mechanisms – other than that proposed by Farley (1963) – contribute to the irregularity structure in the auroral ionosphere.

Acknowledgments

The new research results presented here were obtained through the efforts of many individuals. A much more detailed account of this work, including a description of data analysis procedures, computer programs, radar operating characteristics, radar operating schedules, and so forth is presented in Chesnut et al. (1968). Mr. James C. Hodges of Stanford Research Institute worked very closely with the author on every step of the data analysis. He was also primarily responsible for operation of the Homer radar observatory. Mr. Ray L. Leadabrand provided able leadership, direction, and consultation throughout the program. He was also the motivating force behind the establishment of the Homer radar observatory.

Other members of the Radio Physics Laboratory have contributed to many facets of this work. Mr. Murray J. Baron and Mr. Arvid Larson were responsible for most of the computer programming that was necessary for the proper reduction of the large quantities of data obtained at Homer. Mr. Ronald Presnell has played a major role in the design, assembly, and initial trouble-shooting of the Homer radar. We especially appreciate the advice that Dr. H. F. (Skip) Bates provided on many occasions.

References

Bates, H. F., Belon, A. E., Romick, G. J., and Stringer, W.R.: 1966, 'On the Correlation of Optical and Radio Auroras', *J. Atmos. Terrest. Phys.* **28**, 439–446.

Booker, H. F.: 1956, 'A Theory of Scattering of Non-Isotropic Irregularities with Application to Radar Reflections from the Aurora', *J. Atmos. Terrest. Phys.* **8**, 204.

Bowles, K. L.: 1955, 'Some Recent Experiments with VHF Radio Echoes from Aurora and Their Possible Significance in the Theory of Magnetic Storms and Auroras', Research Report EE248, Technical Report No. 22, Cornell University.

Bowles, K. L., Balsley, B. B., and Cohen, R.: 1963, 'Field-Aligned E-Region Irregularities identified with Acoustic Plasma Waves', *J. Geophys. Res.* **68**, 2485–2501.

Chapman, S.: 1952, 'The Geometry of Radio Echoes From Aurora', *J. Atmos. Terrest. Phys.* **3**, 1.

Chesnut, W. G., Hodges, J. C., and Leadabrand, R. L.: 1968, 'Auroral Backscatter Wavelength Dependence Studies', Stanford Research Institute, Contract No. AF 30(602)-3734, RADC-TR-68-286.

Farley, Jr., D. T.: 1963, 'A Plasma Instability Resulting in Field-Aligned Irregularities in the Ionosphere', *J. Geophys. Res.* **68**, 6083–6097.

Unwin, R. S. and Know, F. B.: 1968, 'The Morphology of the VHF Radio Aurora at Sunspot Maximum – IV', *J. Atmos. Terrest. Phys.* **30**, 25–46. This is the fourth in a series of articles by Knox. This reference lists the previous articles in the series for those interested in pursuing this work.

Vestine, E. H.: 1944, 'The Geographical Incidence of Aurora and Magnetic Disturbances, Northern Hemisphere', *Terrestrial Magnetism and Atmospheric Electricity (J. Geophys. Res.)* **49**, 77.

SCINTILLATIONS OF SATELLITE SIGNALS

L. LISZKA

Kiruna Geophysical Observatory, Kiruna, Sweden

Abstract. The morphology of the scintillation phenomenon is discussed; in particular its geographical distribution, geomagnetic control and diurnal and seasonal variations. Also studies of parameters of ionospheric irregularities are reviewed.

1. Introduction

It has been observed since the earliest satellite observations that under certain conditions satellite signals received on the ground fluctuate in a random manner. This phenomenon has been explained as being due to diffraction of the incident radio waves by the same ionospheric irregularities which are responsible for radio-star scintillation, and is therefore called 'satellite scintillation'. When a plane wave passes through a layer of irregular electron density, the emerging wave shows phase and amplitude variations which reflect electron density fluctuations in the layer. In the case of satellite scintillation the picture is slightly more complicated. Owing to the finite distance between the radio source and the layer the incident wave is no longer plane. Also the rapid motion of a satellite transmitter is a complicating factor. An example of amplitude recording of the ionospheric beacon satellite BE-B at 40 MHz showing heavy scintillation is given in Figure 1.

23 SEPTEMBER 1968

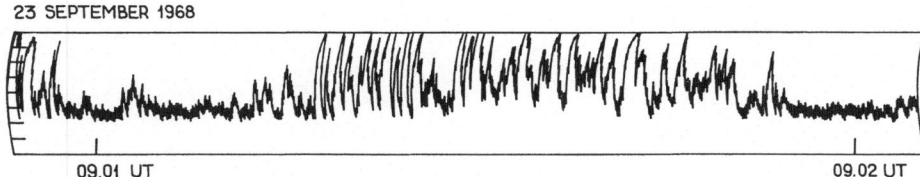

09.01 UT   09.02 UT

Fig. 1. An example of amplitude recording of the ionospheric beacon satellite BE-B at 40 MHz showing heavy scintillation. Recording made at Kiruna Geophysical Observatory.

Scintillation is usually described by the following two parameters:

(1) The depth of the scintillation. This is measured by a 'scintillation index', usually defined as the mean variation in received power relative to the average power. Another definition, using the received amplitude instead of power, is also frequently used.

(2) The scintillation rate. In the case of satellite scintillation it is a function of satellite height and velocity and the irregularity layer parameters: size, elongation, orientation and drift of irregularities.

From a practical point of view the scintillation phenomenon has important consequences for satellite communication. It also gives a simple way of studying ionospheric

D'Angelo (ed.), Low-Frequency Waves and Irregularities in the Ionosphere. All rights reserved

irregularities. The large number of papers which have appeared during the last 10 years on this subject may thus be divided into two groups:

(1) Morphological study of the scintillation phenomenon, often with practical applications.

(2) Study of ionospheric irregularities based on observations of satellite scintillation.

In the present paper results of satellite scintillation studies are reviewed with emphasis on those made at high latitudes.

2. Morphology of the Scintillation Phenomenon

From the point of view of people working with satellite applications such as communication or navigation, the scintillation phenomenon is a serious disturbance especially at high latitudes. An engineer planning a navigation system must know and take into account the probability of occurrence of scintillation as a function of geographical location and geophysical phenomena. This was one of the reasons for the large number of morphological studies of the scintillation phenomenon.

Such aspects as geographical distribution, geomagnetic control of the scintillation phenomenon, its diurnal and seasonal variations and correlation with ionosonde measurements were investigated. The knowledge of these properties of the scintillation phenomenon is also important for theories explaining the phenomenon.

A. GEOGRAPHICAL DISTRIBUTION OF THE SCINTILLATION

The latitude dependence of satellite scintillation was already evident in the observations of early workers; e.g. Kent (1959) recorded 40 MHz signals from the first Russian satellite at Cambridge, and found that in the period of observation scintillation never occurred when the satellite was to the South of Cambridge. His conclusion was that the irregularities producing the scintillation only occurred to the North of latitude ≈ 50° N. The latitude dependence was later studied by a number of authors at different geographical locations and during different periods of the solar cycle. Usually, an increase of scintillation with latitude was found (cf. e.g. Aarons et al., 1963; Liszka, 1963a; Shmelovsky, 1963), although in some cases at subauroral latitudes no definite latitude dependence was found (Frihagen and Tröim, 1961). A detailed comparison of different scintillation measurements is difficult owing to the great diversity of methods of analysis used. Important contributions have been made by a cooperating group of European and American observatories (Joint Satellite Studies Group, 1965, 1968) using a uniform method of analysis of satellite recordings collected during a few periods over a wide range of latitudes. A steady increase of the scintillation index towards the North has been found above Europe over a latitude range from 35°N to 65°N. Figure 2 shows the main result of this work: latitude variation of scintillation index over western Europe in 1962–63 derived from simultaneous overhead observations, together with off-vertical observations showing effect of elevation and propagation angle. Differences between off-vertical observations seem to indicate a latitude variation in the parameters of the scintillation-producing layers, such as layer thickness

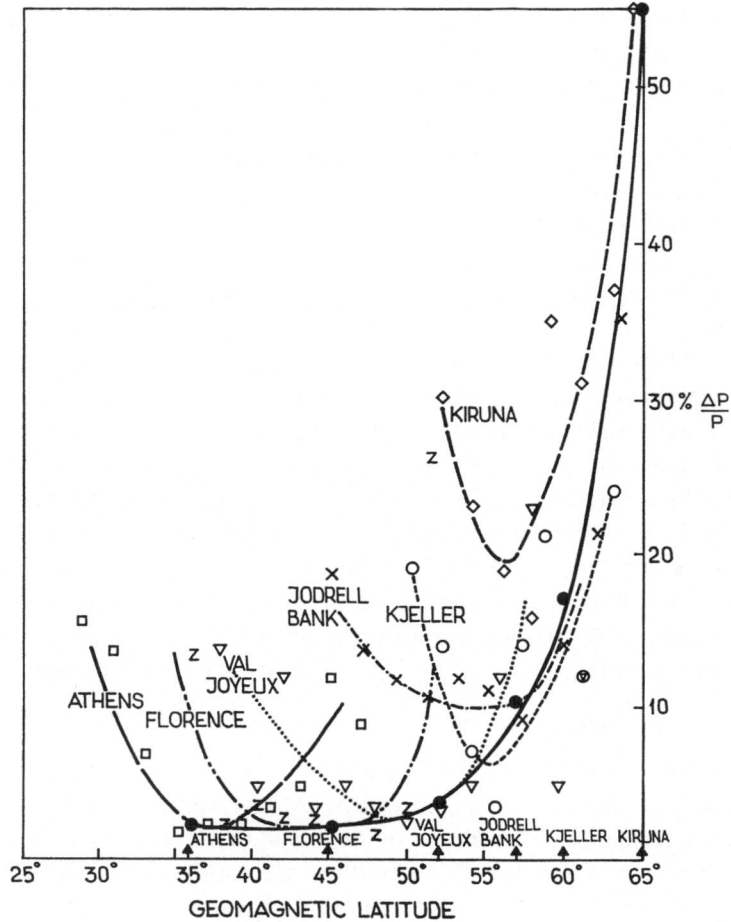

Fig. 2. Latitude variation of scintillation index over Western Europe in 1962–63 derived from simultaneous overhead observations, together with off-vertical observations showing effect of elevation angle (after Joint Satellite Studies Group, 1968).

and average shape of irregularities. Observations from the Southernmost station, Athens (38° N), show an increase of scintillation towards the South. This is probably due to the fact that from a certain latitude the scintillation index starts to increase and reaches a maximum near the geomagnetic equator (Kelleher and Sinclair, 1968). Also top-side soundings have shown (Calvert and Schmid, 1964) that the occurrence and strength of ionospheric irregularities increase in the equatorial regions.

Analysis of the two-dimensional geographic dependence of single station observations (Liszka, 1964b; Whitney et al., 1967) has suggested that contours of constant depth of scintillation tend to align along a latitude dependent on the earth's magnetic field (dip, geomagnetic or invariant latitude). Figure 3 shows contours of constant scintillation index above Scandinavia together with constant-L shells. Observed values of the scintillation index were corrected for the zenith angle dependence by multiplying

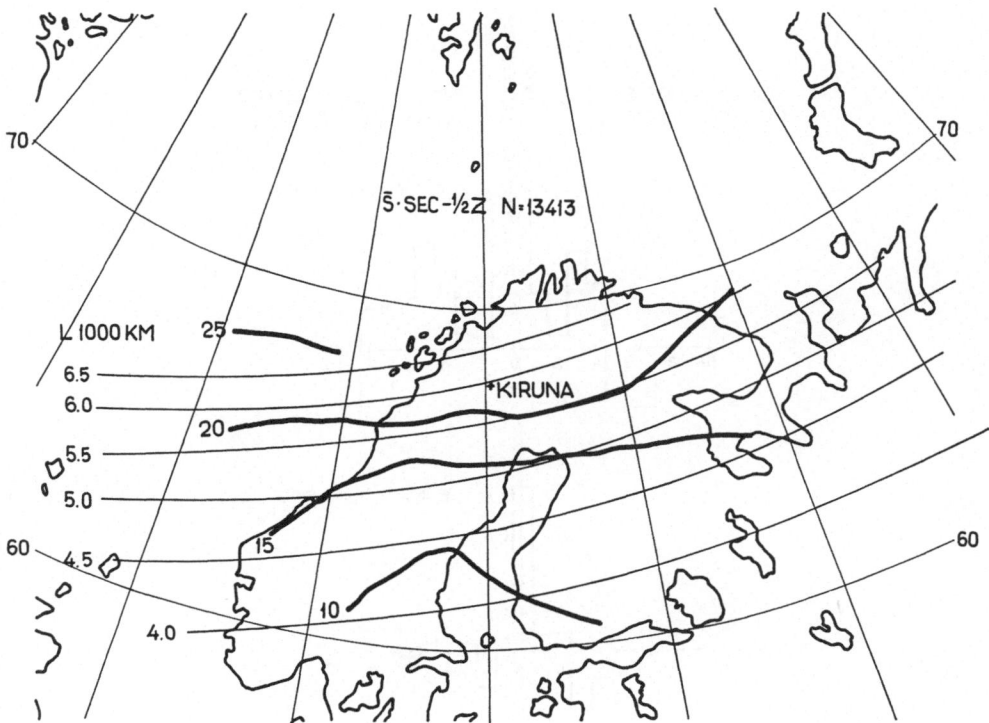

Fig. 3.   Contours of constant scintillation index above Scandinavia together with constant-L shells.

by $(\sec z)^{-1/2}$ (cf. Briggs and Parkin, 1963). A zenith angle dependence of this type seems to be absent at very high latitudes, North of the auroral zone, according to results from Spitsbergen (78° N) (Frihagen, 1968).

An interesting feature which may influence studies of geographic distribution is a peak of the scintillation depth and rate observed at the local geomagnetic zenith, reported by Briggs and Parkin (1963). They have explained it as an effect of end-on looking on ionospheric irregularities elongated along the geomagnetic field lines. The phenomenon, however, is not very distinct (cf. Briggs and Parkin, 1963, Figure 1) and its existence is not confirmed by all results. There is no trace of such a phenomenon at Kiruna (cf. Figure 3) while it is observed by Frihagen (1968) on Spitsbergen or at medium latitudes by Maude and Premanik (1967).

B. GEOMAGNETIC CONTROL OF THE SCINTILLATION

Early work on radio star scintillation has shown (cf. e.g. Dagg, 1957) that the occurrence and the strength of scintillation depend on geomagnetic activity. Subsequently Little *et al.* (1962) found from observations of radio stars carried out in Alaska that the scintillation is correlated with magnetic activity and the lead-lag correlation coefficient between scintillation index and magnetic activity has a maximum for zero time displacement. Similar results, given in Figure 4, have been obtained

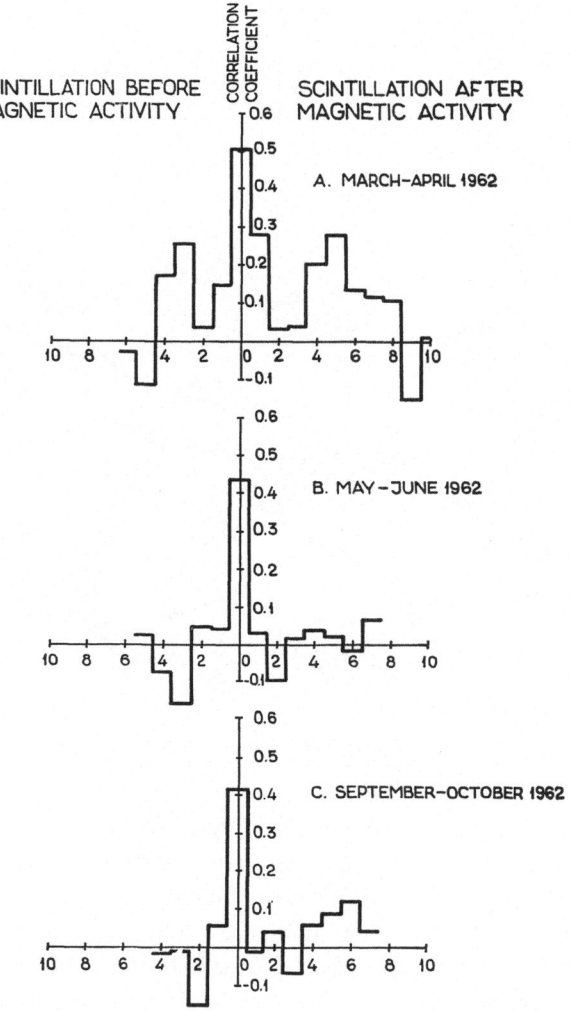

Fig. 4.   Lead-lag correlation coefficients between daily averages of scintillation index in Kiruna
and magnetic Ap-index.

from observations of satellite scintillation at 54 MHz in Kiruna (Liszka, 1963a). At
the same station, a linear dependence between the scintillation index and the local
K-index, with the slope varying throughout the day, has been found (Liszka, 1964a).

Aarons *et al.* (1963) have investigated the geographical distribution of scintillation
as a function of geomagnetic activity. It has been found that at the edge of the auroral
zone the scintillation activity increases and reaches its maximum in the auroral zone.
When geomagnetic activity increases, the 'scintillation zone' moves South (see Figure 5).
Thus at a station located at subauroral latitude, an increase of overhead scintillation
will be observed owing to the motion of the scintillation zone. According to the same
authors, inside the auroral zone there is no apparent change in the scintillation when

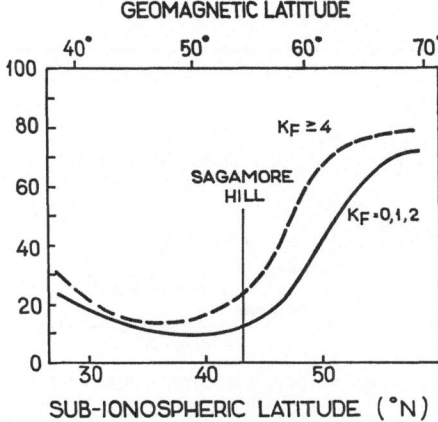

Fig. 5. Geographical distribution of scintillation as a function of geomagnetic activity (after Aarons *et al.*, 1963).

the geomagnetic activity increases. In the equatorial region the variation of scintillation with geomagnetic activity is quite well established (Koster and Wright, 1960). During years near sunspot maximum, there is a strong negative correlation between magnetic disturbances and the occurrence of scintillation. Around years of low solar activity the correlation is negligible.

Also for the scintillation rate a positive correlation with geomagnetic activity has been found (Allen *et al.*, 1964). This may be caused by several factors: a decrease of the scale size of the diffraction pattern over the ground, an increase of the drift velocity, or a more rapid variation of the diffraction pattern. Frihagen (1968) has found on Spitsbergen a linear dependence between the scintillation rate and the scintillation depth. Results obtained in Kiruna (Liszka, 1964c) indicate that the spectrum of the scintillation rate varies with geomagnetic activity and local time. This can probably be explained by local time variation of the production mechanism for the ionospheric irregularities.

C. DIURNAL AND SEASONAL VARIATIONS OF THE SCINTILLATION PHENOMENON

It is well known that the scintillation phenomenon also depends on the time of observation. It has been found from observations of radio stars that scintillation is most frequent during the night. Observations of Cassiopeia made by Dagg (1957) and Chivers (1960) in England have shown that scintillation has a maximum amplitude and rate before local midnight but close to local magnetic midnight. In College, Alaska, it has been found by Little *et al.* (1962) that maximum scintillation occurs at about 0150 local standard time, which corresponds to local magnetic midnight. A secondary maximum at midday has been noted in Australia and Canada but not in England (Booker, 1958).

In a study of satellite scintillation made by Yeh and Swenson (1959) at Urbana, Illinois, it has been shown that the nighttime maximum occurs around local midnight throughout the year. A daytime maximum has also been observed during summer and

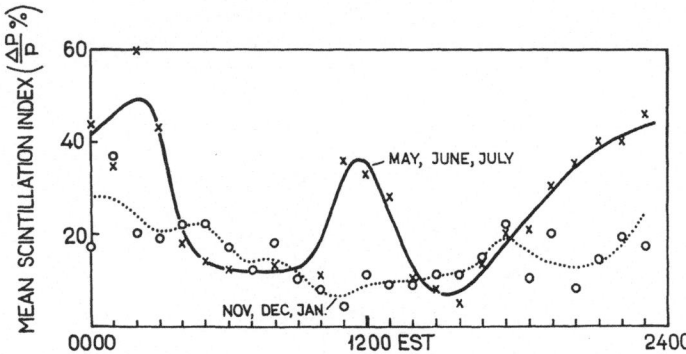

Fig. 6.   Diurnal curves of scintillation index for summer and winter periods at midlatitudes (after
Allen and Mullen, 1966).

autumn months. Also observations of Transit 4A at 54 MHz made at Sagamore Hill,
Mass., during the period July 1961–December 1963 (Allen and Mullen, 1966) show
the same type of diurnal and seasonal variations. Diurnal curves obtained by these
authors for summer and winter periods are shown in Figure 6 where a clear pre-noon
maximum may be seen on the summer curve.

In the equatorial region (Koster, 1968) only one type of diurnal variation with a
nighttime maximum occurs throughout the year. The occurrence of scintillation of
moderate amplitudes centers around local midnight at Legon, Ghana. But the most
violent scintillation (scintillation index greater than 90%) has a peak around 22 hours
local time. The amplitude of the diurnal variation shows a seasonal variation with a
large peak lagging about a month behind the autumn equinox and a smaller peak
near or somewhat later than the vernal equinox.

An investigation of the occurrence of satellite scintillation at 20 MHz in the auroral
zone (Liszka and Hultqvist, 1961) has shown no clear diurnal variation in the
scintillation index. The average level of the scintillation index was very high at night
and only slightly lower during the daytime. However, the results of later observations
at 54 MHz (Liszka, 1963a) have shown a clear diurnal variation in the scintillation
index. A maximum has been found at about 2300 local time, which corresponds well
to local magnetic midnight. No clear seasonal variations have been found.

A reanalysis (Allen and Aarons, 1968) of previous studies made over a wide range of
latitudes by the Joint Satellite Studies Group shows that the amplitude of diurnal
variations is largest at the equator (perhaps more than 20:1) and decreases with latitude. Also the amplitude of seasonal variations was found to decrease with latitude. The
summertime pre-noon maximum was found to be solely a mid-latitude phenomenon.

3. Study of Ionospheric Irregularities Based on Observations of the Satellite Scintillation

When radio waves pass through an irregular ionospheric layer, a diffraction pattern
is produced on the earth's surface. The pattern is usually not stationary, e.g., it may

change in a random way, or it may drift without any change of structure, or it may do both simultaneously. The result of all these factors is that a receiver on the ground will observe amplitude scintillation. For ionospheric irregularities it is often possible to assume (cf. e.g. Ratcliffe, 1956, p. 242) that the pattern does not change during a short period of time, and that therefore only the drift of the pattern is important for the scintillation phenomenon. The drift of the pattern may be caused by the systematic motion of the diffracting screen or by the motion of the source of radio waves. The first is important in the case of radio star scintillation and the second in the case of a satellite transmitter travelling rapidly above the ionosphere.

The pattern has a scale size which is related to the scale size in the ionosphere and moves in the direction of the systematic ionospheric drift, in the case of radio star scintillation, and in the direction opposite to that of the satellite in the second case. One may therefore expect that signals received at ground level will scintillate at a rate proportional to both the scale size of the diffraction pattern and the pattern velocity. It is for this reason that investigations of such patterns may provide information concerning properties of the irregular ionospheric layer. This method of investigating ionospheric irregularities was first used by Mitra in 1949, and by Briggs *et al.* in 1950, for radio star scintillation. The method, based on measurements of the amplitude of the pattern at three spaced receivers, should give the velocity and structure of the pattern. It has, however, been found that the observed pattern is usually anisometric, i.e. has different statistical properties in different directions, and therefore the pattern velocity cannot be found by simple reduction of the time delay between reception at different antennae.

Phillips and Spencer (1955) proposed the characteristic ellipse as a measure of the anisometric properties of the pattern. They assumed that the spatial autocorrelation function in the pattern is constant along an ellipse which therefore represents an average amplitude irregularity on the ground.

Results about the shape and size of the pattern could be used for determining the properties of the ionospheric irregularities using the theory of ionospheric diffraction developed by Booker *et al.* in 1950. Results obtained by radio star scintillation studies have shown that irregularities are situated in and above the *F*-region and are elongated along the geomagnetic field lines.

A. HEIGHT OF IRREGULARITIES

The pattern produced by a satellite is especially suitable for measurements of its velocity and structure, as the direction of the velocity may be known from the knowledge of the orbit. Two antennae situated along the direction of pattern velocity will indicate the time delay between details of the pattern, which will be directly proportional to the pattern velocity. This method, known as 'spaced receiver technique', was first used by Frihagen and Tröim (1961). The principle of this method is shown in Figure 7. The measurements, carried out mostly in Southern Norway, have given results similar to those obtained from radio star scintillation studies. Heights around 350 km have been obtained together with some *E*-region heights.

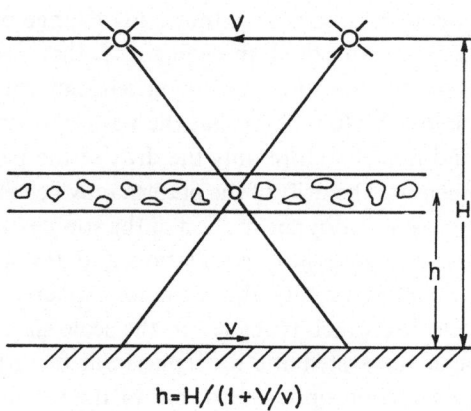

Fig. 7.   Principle of measurement of the height of scintillation producing irregularities.

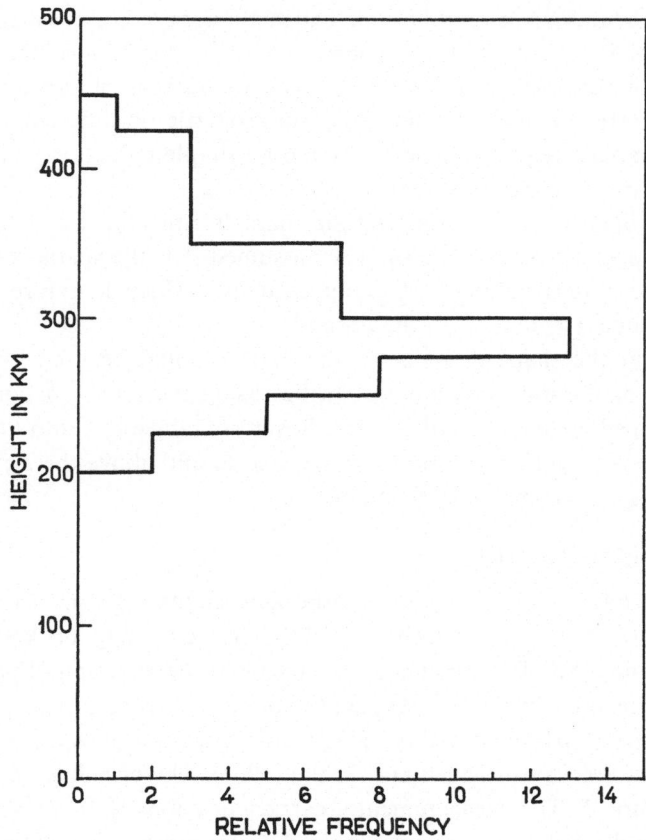

Fig. 8.   Height distribution of scintillation producing irregularities at the equator (after Koster, 1968).

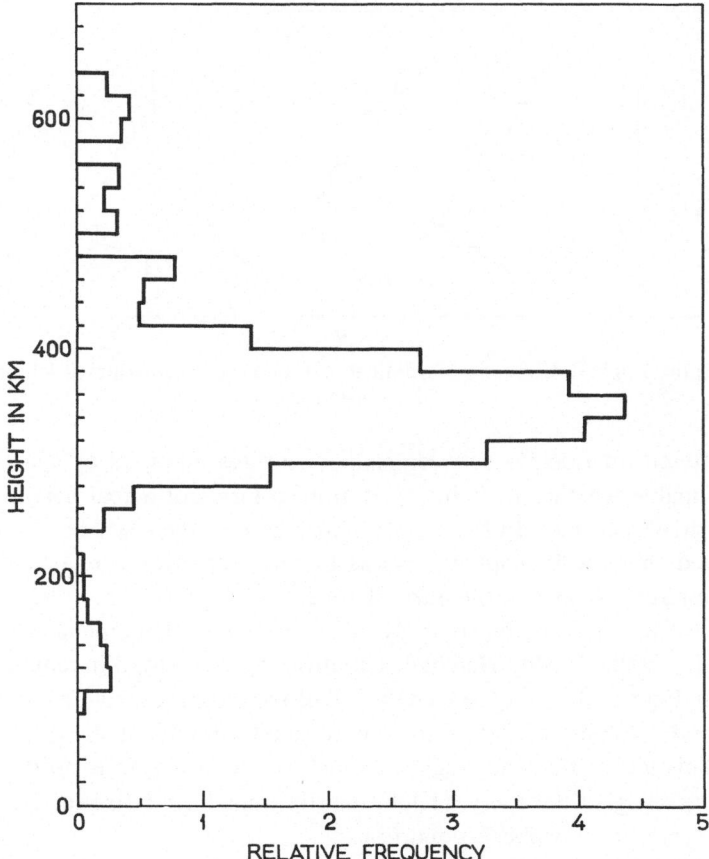

Fig. 9. Height distribution of scintillation producing irregularities at midlatitudes (after Yeh and Swenson, 1966).

A modified spaced receiver technique involving cross-correlation and autocorrelation functions has been introduced by Kent and Koster (1961). From observations of nighttime passes of the satellite 1960 Π 1 (Tiros II) at the equator they obtained heights between 314 and 424 km. Also the latter measurements in the equatorial region (Koster, 1968) have given the height range 200–450 km (see Figure 8).

Also at medium latitudes the predominant part of irregularities seems to be located in the F-region. However, results of Yeh and Swenson (1966), displayed in Figure 9, show some irregularities at E-region heights. This result has not been confirmed by later measurements at the same observatory (Paul, 1968).

At high latitudes irregularities were observed at heights above 100 km (Hook and Owren, 1962; Liszka, 1963b; Liszka, 1964d). The height distribution there is different from that at medium latitudes; irregularities seem to be distributed over a wide height interval from 100 km upwards. Sometimes extremely scattered height values were obtained (Liszka, 1964; Frihagen and Liszka, 1965). Because these are obtained from

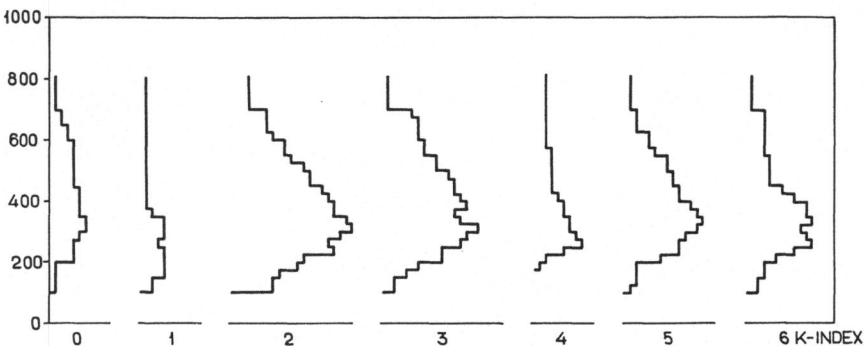

Fig. 10. Relative height distributions of scintillation producing irregularities at Kiruna for different
local K-indices.

widely scattered time delays between similar features observed by spaced antennae,
it is questionable whether their interpretation in terms of a real height distribution
of irregularities is correct. In the simple theory used in the spaced receiver technique
it is assumed that the ionosphere acts as a thin diffraction screen, an assumption
probably not always valid in the auroral zone.

An attempt has been made to study effects of other phenomena on these height
distributions (Liszka, 1964d). Height distributions obtained for different local K-indices
are shown in Figure 10. It may be seen that all distributions, except for the distributions
for $K=0$ and 1 (which include only a very small amount of data), are practically
identical. This means that the height at which the scintillation is produced does not
depend on geomagnetic activity. Height distributions have also been constructed for
3 different types of ionospheric conditions:

(1) only F-layer echoes present on the vertical soundings;

(2) sporadic-E echoes present;

(3) no ionospheric echoes – auroral 'black-out'.

The distributions obtained are shown in Figure 11. It may be seen that the extremely
wide height distribution is associated with cases of auroral blackout. This seems to
confirm the hypothesis that the widely scattered time delays, sometimes observed at
high latitudes, are due to the presence of a thick diffracting screen and may not be
represented by a height distribution according to the simple theory.

Also the latitudinal dependence of the height distribution has been studied. McClure
and Swenson (1964) observed a slight increase of scintillation height towards the
North. Results from Gorkij and Murmansk (Getmantsev et al., 1968) covering the
latitude range 53–77° N give weak evidence of a similar behavior. The Joint Satellite
Studies Group made simultaneous spaced receiver measurements at several observatories (Bratteng and Frihagen, 1966; Liszka, 1968). The results obtained do not show
any definite variations of height with geographical location. It has been found that
the variations of time delays observed during different passes support the assumption
used in both studies that the ionospheric irregularities are elongated along the geomagnetic field lines.

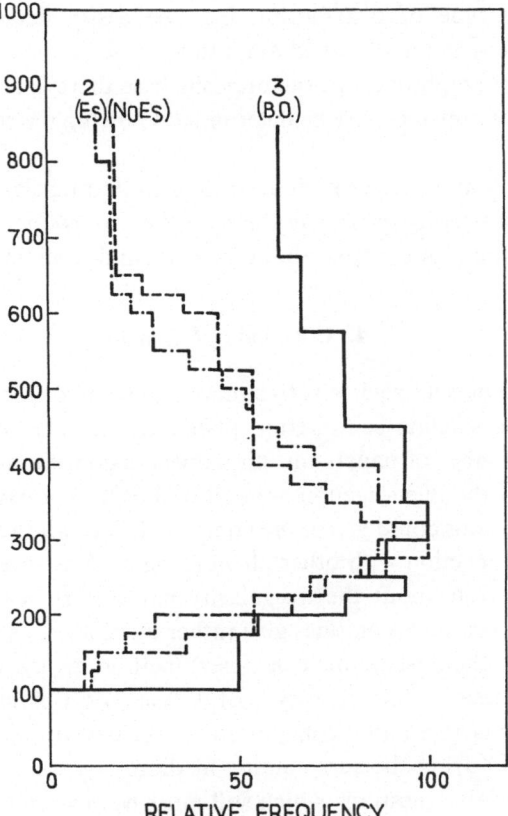

Fig. 11.   Relative height distributions of irregularities occurring during
different ionospheric conditions.

B. AMPLITUDE DISTRIBUTIONS

It may be assumed from the theory of diffraction (Ratcliffe, 1956) that the amplitude
and phase distributions of the received signal are dependent on the statistical properties of the scattering medium, in particular on the standard deviation, N_m, of the
number density of electrons. In the case of large density fluctuations, there will be a
Rayleigh type amplitude distribution; for small values of N_m, a displaced Gaussian
distribution is obtained. Rice (1944) introduced a probability distribution function
which includes both distributions, together with all intermediate cases.

It has been shown (Wernik and Liszka, 1968) that amplitude distributions of
scintillating satellite signals do not agree with the theoretical Rice distributions. The
observed distributions have been explained assuming a correlation between components of the field vector of the scattered wave. This means that even a strongly
scattered wave is in such a case polarized with a degree of polarization determined
by the correlation coefficient between components of the field vector. The above-
mentioned assumption leads to a Wheelon type amplitude distribution (Wheelon, 1957).

In practice, the type of distribution may be easily determined using a ratio $K = \bar{A}^2/(\bar{A})^2$, where $\bar{A}$ is the observed amplitude. If K is smaller than $4/\pi$, there is a Rice distribution of amplitudes. In the opposite case there is a Wheelon distribution. Also distribution parameters may be determined knowing the ratio K (cf. Wernik and Liszka, 1968).

The knowledge of amplitude distribution of scintillating signals gives useful information about the scattering process in the scintillation-producing layer and indirectly about the irregularities, for example about their size and sometimes about the thickness of the layer.

4. Concluding Remarks

The purpose of the present review is to summarize some of the observational results from studies of the satellite scintillation phenomenon. The review is certainly not complete, as the number of papers on this subject is continuously increasing.

Future studies of the phenomenon are still desirable. For example, well organized world-wide observations could give more complete information about its geographical distribution and connection with other phenomena such as spread-F and sporadic-E. Also more information about the irregularity parameters is needed. In particular, the present spaced receiver techniques give rather unreliable results, especially at high latitudes. Therefore the development of a new method giving accurate height distribution of irregularities would be very useful. Very little is still known about the dimensions of irregularities and the magnitude of the deviation of the electron density within an irregularity from the surrounding medium.

These are some of the questions which still must be answered for a more complete understanding of the processes responsible for the production of ionospheric irregularities.

References

Aarons, J., Mullen, J., and Sunanda Basu: 1963, 'Geomagnetic Control of Satellite Scintillations', *J. Geoph. Res.* **68**, 3159.

Allen, R. S. and Aarons, J.: 1968, 'Analysis of Special Overhead Observations: II. Diurnal Variation', JSSG Rep. No. 3, 96.

Allen, R. S., Aarons, J., and Whitney, H.: 1964, *Trans. IEEE Ant. Prop.*, AP-12, 812.

Allen, R. S. and Mullen, J. P.: 1966, 'A Diurnal and Seasonal Analysis of Scintillations of Transit 4A', JSSG Rep. No. 2, 102.

Booker, H. G.: 1958, 'The Use of Radio Stars to study Irregular Refraction of Radio Waves in the Ionosphere', *Proc. IRE* **46**, 298–314, 1085.

Booker, H. G., Ratcliffe, J. A., and Shinn, D. H.: 1950, 'Diffraction from an Irregular Screen with Applications to Ionospheric Problems', *Phil. Trans. Roy. Soc. London* **242**, 579–607.

Bratteng, O. and Frihagen, J.: 1966, 'Survey of Ionospheric Irregularities over Europe', FFIE Rep. E-83.

Briggs, B. H. and Parkin, I. A.: 1963, 'On the Variation of Radio Star and Satellite Scintillations with Zenith Angle', *J. Atm. Terr. Phys.* **25**, 339.

Briggs, B. H., Phillips, G. J., and Shinn, D. H.: 1950, 'Analysis of Radio Fading at Spaced Receivers', *Proc. Phys. Soc.* **B63**, 106–121.

Calvert, W. and Schmid, C. E.: 1964, 'Spread-*F* Observations by the Alouette Topside Sounder Satellite', *J. Geophys. Res.* **69**, 1839–1852.

Chivers, H. J. A.: 1960, 'Observed Variations in the Amplitude Scintillations of Cassiopeia (23N5A) Radio Source', *J. Atm. Terr. Phys.* **19**, 54–64.

Dagg, M.: 1957, 'The Origin of the Ionospheric Irregularities Responsible for Radio-Star Scintillations and Spread-F', *J. Atm. Terr. Phys.* **11**, 133–150.

Frihagen, J.: 1968, 'Satellite Scintillation at High Latitudes and Its Possible Relation to Precipitation of Soft Particles', JSSG Rep. No. 3, 73.

Frihagen, J. and Liszka, L.: 1965, 'A Study of Auroral Zone Ionospheric Irregularities made simultaneously at Tromsö, Norway, and Kiruna, Sweden', *J.A.T.P.* **27**, 513.

Frihagen, J. and Tröim, J.: 1961, 'On the Large-Scale Regions of Irregularities producing Scintillation of Signals transmitted from Earth Satellites', *J. Atm. Terr. Phys.* **20**, 215.

Getmantsev, G. G., Gringauz, K. I., Eroukhimov, L. M., Kravtsov, Yu. A., Mityakov, N. A., Mityakova, E. E., Roudakov, V. A., and Rytov, S. M.: 1968, 'Investigation of Electron Density in the Ionosphere by Means of the Ground-Based Observations of Radio Signals radiated from Cosmic Vehicles (a Review)', *Radiophys.* **11**, 649.

Hook, J. L. and Owren, L.: 1962, 'The Vertical Distribution of E-Region Irregularities deduced from Scintillations of Satellite Radio Signals', *J. Geophys. Res.* **67**, 5353.

Joint Satellite Studies Group: 1965, 'A Synoptic Study of Scintillations of Ionospheric Origin in Satellite Signals', *Planetary Space Sci.* **13**, 51–62.

Joint Satellite Studies Group: 1968, 'On the Latitude Variations of Ionospheric Origin in Satellite Signals', *Planetary Space Sci.* **16**, 775.

Kelleher, R. F. and Sinclair, J.: 1968, 'Diurnal and Latitudinal Dependence of Scintillations in Satellite Transmissions', JSSG Rep. No. 3, 98.

Kent, G. S.: 1959, 'High-frequency Fading observed on the 40 Mc/s Wave radiated from Artificial Satellite 1957a', *J. Atm. Terr. Phys.* **16**, 10.

Kent, G. S. and Koster, J. R.: 1961, 'Height of Nighttime F-Layer Irregularities at the Equator', *Nature* **191**, 1083.

Koster, J. R.: in *Proceedings of NATO Advanced Study Institute on Satellite Signal Propagation in the Ionosphere, Viareggio, 10–22 June 1968* (in press).

Koster, J. R. and Wright, R. W.: 1960, *J. Geoph. Res.* **65**, 2303.

Liszka, L.: 1963a, 'Satellite Scintillation observed in the Auroral Zone', *Arkiv Geofys.* **4**, 211.

Liszka, L.: 1963b, 'A Study of Ionospheric Irregularities using Satellite Transmissions at 54 Mc/s', *Arkiv Geofys.* **4**, 227.

Liszka, L.: 1964a, 'Dependence of Auroral Zone Satellite Scintillations on Geomagnetic Activity', *Arkiv Geofys.* **4**, 445.

Liszka, L.: 1964b, 'The Geographical Distribution of Satellite Scintillation Activity above Scandinavia', *Arkiv Geofys.* **4**, 453.

Liszka, L.: 1964c, 'A Study of the Scintillation Rate Spectrum observed in the Auroral Zone', *Arkiv Geofys.* **4**, 529.

Liszka, L.: 1964d, 'An Investigation of the Height of Scintillation producing Irregularities', *Arkiv Geofys.* **4**, 523.

Liszka, L.: 1968, 'JSSG-Spaced Receiver Measurements', Final Scientific Report, Contract F 61052 67 C 0072.

Liszka, L. and Hultqvist, B.: 1962, 'Investigations of Radio Transmissions from 1958 (Sputnik III) made at Kiruna Geophysical Observatory', *Arkiv Geofys.* **4**.

Little, C. G., Reid, G. C., Stiltner, E., and Merritt, R. P.: 1962, 'An Experimental Investigation of the Scintillation of Radio Stars observed at Frequencies of 223 and 456 Megacycles per Second from a Location Close to the Auroral Zone', *J. Geophys. Res.* **67**, 1763–1783.

McClure, J. P. and Swenson, G. W., Jr.: 1964, 'Beacon Satellite Studies of Small-Scale Ionospheric Inhomogeneities', Univ. Illinois Dept. Electr. Eng., Urbana, Ill., Techn. Rep.

Maude, A. D. and Premanik, M. A.: 1967, *J.A.T.P.* **29**, 1311.

Mitra, S. N.: 1949, *I.E.E.* **96**, 441.

Paul, L. M.: 1968, 'Measurement of Irregularity Heights by the Spaced Receiver Technique', Univ. Illinois Dept. Electr. Eng., Urbana, Ill., Techn. Rep. No. 36.

Phillips, G. J. and Spencer, M.: 1955, 'The Effects of Anisometric Amplitude Patterns in the Measurement of Ionospheric Drifts', *Proc. Phys. Soc.* **B68**, 481–492.

Ratcliffe, J. A.: 1956, 'Some Aspects of Diffraction Theory and Their Applications to the Ionosphere', *Rep. Progr. Phys.* **XIX**, 188.

Rice, S. O.: 1944, 'Mathematical Analysis of Random Noise', *Bell System Tech. J.* **23**, 282.

Shmelovsky, K. H.: 1963, 'On the Latitudinal Dependence of the Fluctuation Index of the Satellite Wireless Signals on 20 Mc/s', *Geomagnetism and Aeronomy* **3**, 1129.

Wernik, A. W. and Liszka, L.: 1968, 'On the Amplitude Distribution of Scintillating Radio Signals from Artificial Satellites', *Arkiv Geofys.* **5**, 501.

Wheelon, A. D.: 1957, 'Relation of Radio Measurements to the Spectrum of Tropospheric Dielectric Fluctuations', *J. Appl. Phys.* **28**, 684.

Whitney, H. E., Allen, R. S., and Aarons, J.: 1967, 'Studies of the Latitudinal Variations of Irregularities by Means of Synchronous and 1000 km Satellites', in *Space Research* **VII**, North-Holland Publ. Co., pp. 1358–1369.

Yeh, K. C. and Swenson, G. W., Jr.: 1959, 'The Scintillation of Radio Signals from Satellites', *J. Geophys. Res.* **64**, 2281–2286.

Yeh, K. C. and Swenson, G. W., Jr.: 1966, '*F*-Region Irregularities studied by Scintillation of Signals, Spread-*F* and Its Effects upon Radiowave Propagation and Communications', AGARDograph 95.

INHOMOGENEITIES IN THE IONOSPHERE MEASURED BY RADIO SIGNALS FROM THE BEACON SATELLITE EXPLORER-22, EMPHASIZING SATELLITE SCINTILLATIONS

G. K. HARTMANN

Max-Planck-Institut für Aeronomie, Lindau/Harz, Germany

1. Introduction

Since November 1964 the amplitude of radio signals from the beacon satellite Explorer-22 has been recorded in Lindau for the purpose of obtaining the ionospheric electron content from the Faraday effect [1] and the differential Doppler effect [2]. The frequencies used were 20 MHz, 40 MHz, 41 MHz, 136 MHz and 360 MHz. On a considerable number of occasions the regular Faraday-fading and Doppler-fading effects were distorted or even obscured by other effects. Most of these were due to inhomogeneities in the ionosphere, a few of them ($\sim 7\%$) to inhomogeneities in the troposphere [3]. The recordings obtained in Lindau showed two types of 'other' effects, (a) so-called satellite scintillations and (b) effects which were the result of horizontal gradients in the ionosphere. Horizontal gradients are defined as variations in the electron content N within the ionosphere at a constant height h along a path S ($dN/dS \neq 0$).

Here we shall consider satellite scintillations and discuss the results in detail. We shall also give a few examples of the other type.

About 25% of all the recordings obtained in Lindau showed scintillations. These were correlated as far as possible with relevant ionograms and in addition with

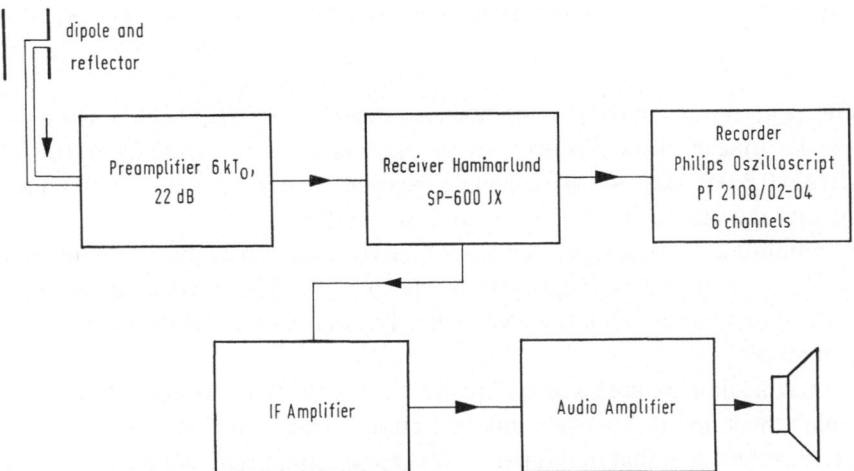

Fig. 1.   Block diagram of the Faraday-effect receiving system.

Kp values. Furthermore their duration, their time of occurrence, and the azimuth range of the occurrence were examined.

Because we have more Faraday recordings than differential Doppler recordings, we shall give the results of the reduced Faraday data.

2. Measuring Technique

Figure 1 is the block diagram of the Faraday-effect receiving system. Recently a much better and more sensitive so-called 'polarimeter system' has become available. But this is much more expensive than the equipment shown in Figure 1.

Figure 2 is a regular recording of the amplitude of the radio signals from the

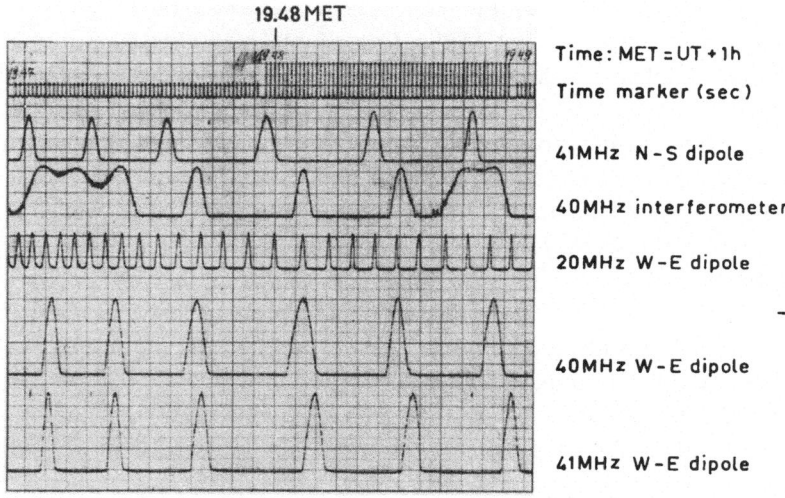

Fig. 2.   Faraday-effect recording, Satellite Explorer-22 (S-66), July 6, 1965. Rev. No. 3705.

satellite Explorer-22 covering a 3-min period during its passage. The uppermost trace shows 1-sec time markers. The other traces, reading downwards, are 41 MHz, 40 MHz, 20 MHz, 40 MHz, and 41 MHz signals received on various antenna systems. E–W means that the dipole axis was directed East–West, etc.

The amplitude of each signal increases linearly downwards (but is limited at about 20 dB above rise) and time goes from left to right. This recording was made on July 6th, 1965, during revolution No. 3705. The paper speed of the recording system was 2 mm/sec.

At the same time we used a second paper recorder with a paper speed of 15 mm/min to display the amplitude without any limitation. Figure 3 is a recording made in this way. It must be noted that in this particular case the amplitude of each signal increases upwards and time goes from right to left. A complete passage of the satellite Explorer-22, revolution No. 7262, made on March 22nd, 1966, is displayed.

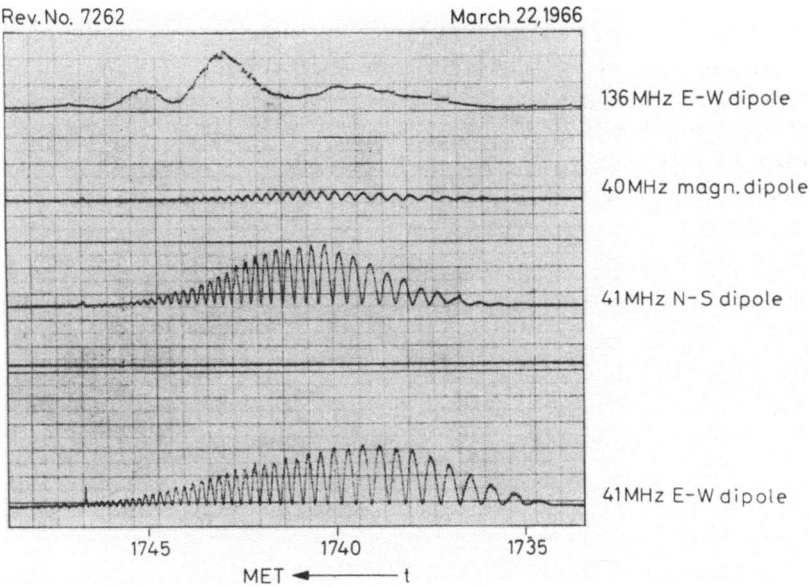

Fig. 3.   Faraday-effect recording, Satellite Explorer-22 (S-66), March 22, 1966. Rev. No. 7262.

3. Satellite Scintillations, Background [6]–[10]

In many cases the usual regular Faraday fading – the duration of the fading periods varies between 1 sec and several minutes – is accompanied by another, very fast and irregular, fading. The duration of these fading periods varies between 2 Hz and about 100 Hz. Figure 4 shows a recording with regular Faraday fading on the left and right, and in the middle ∼2 min of very fast fading. This is known as satellite scintillation. The uppermost trace shows 1-sec time markers. The other traces, reading downwards, are 20 MHz, 41 MHz, 40 MHz, 40 MHz, and 41 MHz. The amplitude increases linearly downwards (but is limited at about 20 dB above rise) and time goes from left

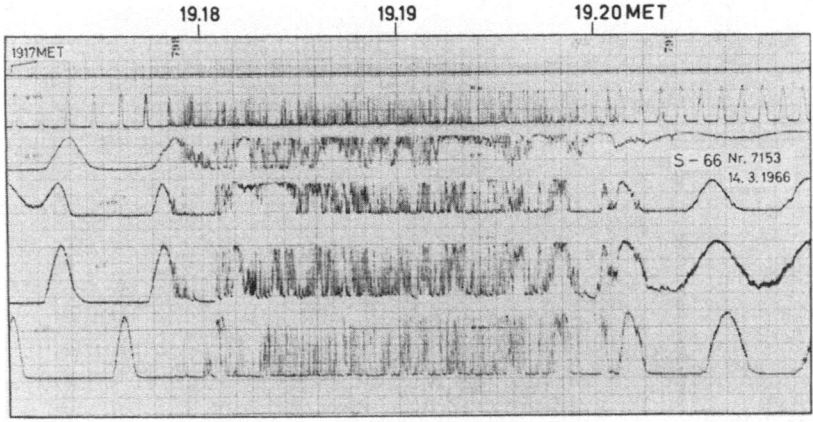

Fig. 4.   Faraday-effect recording with satellite scintillation S-66, March 14, 1966. Rev. No. 7153.

to right. This recording was made on March 3rd, 1966, during revolution No. 7153.

Satellite scintillations are the result of diffraction of radio waves by inhomogeneous structures within the ionosphere. Similar irregular intensity variations are observable in radio signals emitted by cosmic radio sources. It is very likely that these effects are also caused by inhomogeneities within the ionosphere. To this one applies the term 'radio star scintillations'. Several authors have compared the satellite scintillations with the occurrence of radio star scintillation [4].

For the initial purpose of obtaining the total electron content, differential Doppler recordings were made along with the Faraday recordings of signal amplitude. The differential Doppler recordings essentially display the phase difference of two co-herently related signals. These are basically a measure of the ratio of change of phase path between the satellite and receiver, due to the ionosphere.

Figure 5 shows an example of a differential Doppler recording [2] e.g. between the

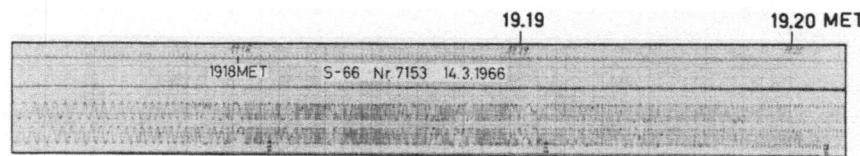

Fig. 5.   Differential Doppler recording with satellite scintillation S-66, March 14, 1966. Rev. No. 7153.

40 MHz and 360 MHz signals of Explorer-22, rev. No. 7153, on March 3rd, 1966. The uppermost trace displays 1-sec time markers. The chart speed was 5 mm per sec. The 3rd and 4th traces show the phase path variation. Using the same satellite signal, trace 3 shows the cosine output of the phase lock loop and trace 4 the sine output. This is to prevent ambiguity when deciding whether the phase path has become longer or shorter. On the left and right of this recording regular phase path variations are to be seen. In the middle – for 2 min – the phase is heavily distorted and shows satellite scintillations. Looking back to the amplitude recording Figure 4, which was recorded at the same time as the 'phase recording' Figure 5, one can compare the effect of satellite scintillations on amplitude and phase recordings.

Considering the ionosphere as a phase screen in a mean ionospheric height one can easily combine the temporal autocorrelation function of the recorded field strength with the spatial autocorrelation function of the diffracting screen [5]. We did not examine our recordings in this way because we did not have additional data to check some of the assumptions which were essential in applying the above-mentioned evaluation method.

All amplitude recordings done with signals from Explorer-22 between November 28th, 1965, and November 30th, 1966, were examined for the occurrence of satellite scintillations. About 1200 recordings were analyzed. In order to exclude spurious effects such as those resulting from interference by signals from broadcasting and other transmitters we have noted for correlation purposes only those occasions on which scintillations were observable in both the 40 MHz and 41 MHz traces. The maximum recording time of this satellite – 80° inclination – was 14 min in Lindau.

4. Results

302 recordings showed satellite scintillations. The above-mentioned inhomogeneities by which the scintillations were caused were located anywhere along the ray path between the satellite and receiver. The known ephemerides of the satellite's orbit enable us to calculate its topocentric coordinates, azimuth and elevation angle. These data could be obtained by means of optics from any receiving station. In order to determine the exact location in space of the inhomogeneities one needs, in addition to the azimuth and elevation angle, exact measurements of the time of a pulse reflected by such inhomogeneities or some additional spaced receiving stations for 'triangulation' purposes.

The 1200 recordings were arranged in three different groups.

I. Type 0: All recordings that showed no scintillation at all. We found 898 of these.

II. Type 1: All recordings that showed weak scintillations. In this particular case we used those recordings where the fast amplitude variations were less than or equal to 3 dB. We found 118 of these.

III. Type 2: All recordings that showed strong scintillations. In this particular case we used those recordings where the fast amplitude variations were greater than 3 dB. We found 184 of these.

We chose the order 0, 1, 2, so that we could give an interpretation of the relative amplitude variations instead of the absolute ones, which are much affected by complicated physical parameters. All those recordings which showed intervals of scintillations more than about 30 sec apart were put into the category to which the longest scintillation interval was logically connected. This classification was not applicable to 8 recordings.

The following section gives a temporal and spatial classification of the measured satellite scintillations.

(a) *Duration of the satellite scintillations*

Duration t	Number of recordings
30 sec $\leq t <$ 1 min	19
1 $\leq t <$ 5 min	174
5 $\leq t <$ 10 min	109
10 $\leq t <$ 14 min	10

(b) *Time of occurrence*

Time within	Number of recordings	
	Type 1	Type 2
1.00 to 9.00 MET	68	41
9.00 to 17.00 MET	15	4
17.00 to 1.00 MET	74	94

This classification was not applicable to 6 recordings.

(c) *Range of azimuth*

	within SW → NW	within NW → NE	within NE → SE	within SE → SW
Number of rec.	60	65	55	49

92 recordings showed an interval of scintillation that extended over more than 90° azimuth.

(d) *Range of elevation angle*

No detailed examination was carried out because there would have been no reasonable physical interpretation.

(e) *Correlation with bottom-side ionosonde data*

(e1) *Comparison with vertical incidence ionograms*

All the recordings were used which showed satellite scintillations very close to the zenith – elevation angle ⩾ 85°. These recordings were compared with relevant ionograms. Table I gives the results.

TABLE I

Revolution No.	Date	Time (MET)	Type of scintillation	Spread F	Es (MHz)	Kp
5918	14.12.65	21.30	2	strong	3.6	1+
6347	15. 1.66	3.20	1	strong	2.4	20
6388	18. 1.66	3.00	1	medium	2.7	0+
7112	11. 3.66	19.35	2	–	–	2 –
7153	14. 3.66	19.20	2	–	2.0	2+
8131	24. 5.66	23.15	2	medium	2.0	10
8295	5. 6.66	21.50	2	medium	3.1	1 –
8765	10. 7.66	3.10	2	strong	4.9	4+
8806	13. 7.66	2.45	2	strong	3.3	10
8847	16. 7.66	2.20	2	strong	7.2	1+
8929	22. 7.66	1.40	2	strong	–	40
9060	31. 7.66	13.50	1	–	7.8	10
9907	1.10.66	5.30	1	medium	3.1	–
10631	23.11.66	22.15	1	very weak	–	–

These 14 recordings represent somewhat more than 1% of the total number of evaluated recordings. If a physical interpretation of such a small percentage were to be allowed we would have to emphasize that no correlation exists with the occurrence of spread F or Es in the ionograms. As is to be expected, spread F with the absence of Es and vice versa causes satellite scintillations. Furthermore it seems to be possible to measure disturbances – along the ray path – above the $F2$ max. This result was obtained during revolutions No. 7112 and No. 10631. During the passage of revolution No. 7112 we found very strong satellite scintillations. The simultaneously recorded ionogram displayed neither spread *F* nor *Es*. During the passage of revolution No. 10631 we found satellite scintillations but we had only very weak spread *F*.

(e2) *Comparison with oblique incidence ionograms*

We were able to use 41 recordings on which scintillations were displayed within a solid angle corresponding approximately to the main lobe of the rhombic antenna used by the ionosonde. In order to carry out the ionosonde and satellite measurements simultaneously, we followed a programme of co-operation between our two field stations from February 1966 to August 1966. The 41 recordings were made during this time. We were able to classify 26 of them as type 1 and 15 as type 2. Almost all the relevant ionograms indicated slight and strong disturbances in the ionosphere.

In 5 of the 41 samples there were no distortions observable in the ionograms. We conclude that the disturbances were located above *F*2 max. Furthermore we compared recordings from type 0 with ionograms in order to determine whether disturbances existed in the ionosphere which could be detected by ionograms but not by a simultaneously made satellite recording. An unambiguous result is not yet available. For this purpose we need detailed measurements of the different lobes of the rhombic antenna. These measurements are to be made.

(f) *Conclusions*

At our medium latitude ($\sim 51°$) about 25% of all satellite recordings showed satellite scintillations. There was no correlation between Kp values and the occurrence of satellite scintillations. Apart from the comparison with these Kp values only 10% of all the satellite scintillations measurements could be compared with available ionosonde and other ionospheric data. So scintillation data from a one-year period do not enable us to give a very detailed physical interpretation but only to study the general features of this phenomenon. It must be emphasized that in the case of satellite scintillations simple and normal models of the ionosphere based on geometric optics are no longer valid. Since wave optics causes the physical effects, it is impossible to apply ray tracing methods.

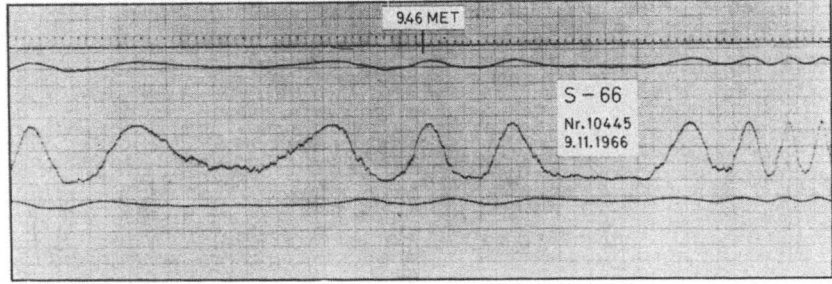

Fig. 6. Faraday-effect recording with horizontal gradients (41 MHz) S-66, November 9, 1966. Rev. No. 10445.

9.46 MET

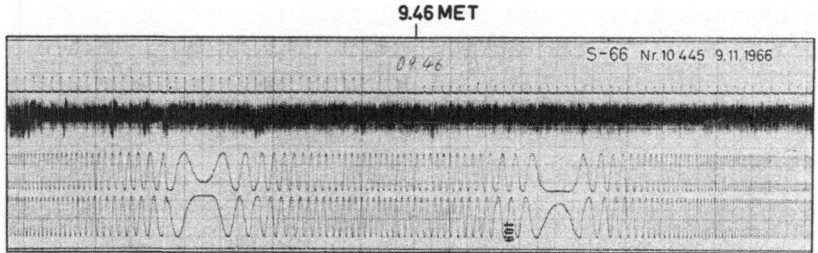

Fig. 7. Differential Doppler recording with horizontal gradients S-66, November 9, 1966.
Rev. No. 10445.

At our medium latitude and with one observing station one can only investigate the phenomenon itself as was done in the previous chapters. No detailed physical interpretation seems to be possible. For a better understanding of this phenomenon one would need additional data, which are scarce as yet. The absolute value of the refractive index was in all these cases >0.9 for the relevant observed f_0F2.

HORIZONTAL GRADIENTS

As we have already mentioned we shall give only a few examples of this type of ionospheric distortion.

Figure 6 is a recording of the 41 MHz radio signal, displayed by a separate paper recorder. We used different antenna systems. The paper speed was 5 mm/sec. The uppermost trace shows 1-sec time markers. The other traces, reading downwards, are the 41 MHz signal using an E–W interferometer, an E–W dipole, and an N–S

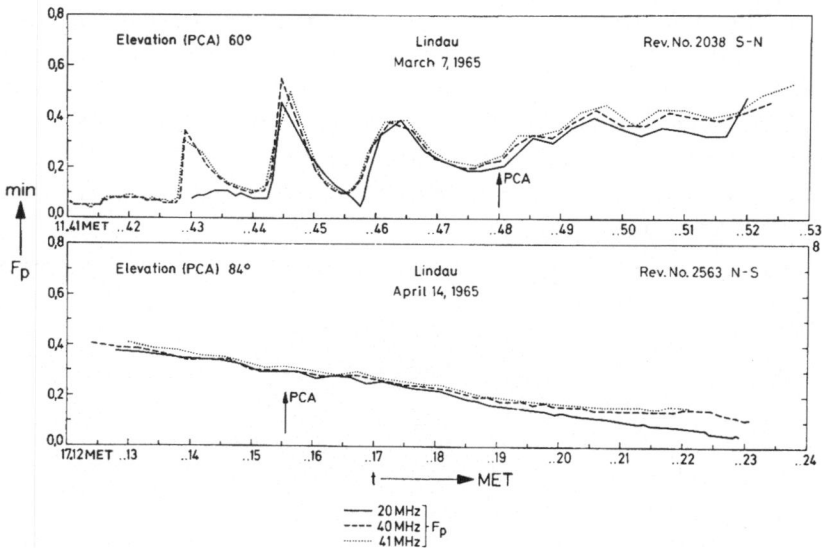

Fig. 8. Two graphic presentations of recordings, Rev. No. 2038 and 2563.
PCA = Point of Closest Approach.

dipole. Time goes from left to right. The amplitude of the signal increases downwards. This recording was made on November 11th, 1966, during revolution No. 10445. Normally the fading periods increase or decrease steadily. In this case we find a very different behaviour. Two long Faraday periods are followed by short ones. These long periods which alternate with short periods within such a short time interval indicate pronounced horizontal gradients in the ionosphere.

Figure 7 shows the relevant differential Doppler recording. For channel distribution see Figure 5. The two long Faraday fading periods correspond very well with the two long periods on the differential Doppler recording. Figure 8 gives a graphic presentation of one recording – revolution No. 2038 – with strong horizontal gradients and of one – revolution No. 2563 – with regular behaviour. The duration of the Faraday fading periods [3] is presented as a function of the time during which the signals were observable. Because of the frequency dependence of the Faraday effect we drew the duration of 4 fading periods for the 20 MHz trace instead of 1 as we did for 40 MHz and 41 MHz. The 20 MHz function is displayed with a solid line, the 40 MHz function with a dashed line and the 41 MHz function with a dotted line.

More than 65% of all recordings showed such horizontal gradients within the ionosphere. In many cases ray tracing methods might be applicable.

Acknowledgements

The research reported here has been supported by the German Ministry of Scientific Research under research grant WRK 125.

References

[1] Hartman, G.: 1965, 'Bestimmung der Elektronendichte zwischen der Erdoberfläche und einem künstlichen Erdsatelliten mit Hilfe des Faraday-Effektes', *AEÜ* **19**, 207–214.

[2] Schmidt, G.: 1966, 'Bestimmung des Elektroneninhaltes zwischen der Erdoberfläche und einem künstlichen Erdsatelliten mit Hilfe des Differenz-Doppler-Effektes', *AEÜ* **20**, 374–378.

[3] Hartmann, G.: 1967, 'Die Amplitudenregistrierungen des Satelliten Explorer-22, unter besonderer Berücksichtigung der Effekte, die bei Elevationswinkeln kleiner als 45° auftreten', *Mitt. Max-Planck-Inst. für Aeronomie*, Nr. 31.

[4] Yeh, K. C. and Swenson, G. W.: 1959, 'The Scintillation of Radio Signals', *JGR* **64**, 2281.

[5] Ratcliffe, J. A.: 1956, 'Some Aspects of Diffraction Theory and Their Application to the Ionosphere', *Rep. Progr. Phys.* **19**, 188–267.

[6] Aarons, J.: 1963, *Radio Astronomical and Satellite Studies of the Atmosphere*, North-Holland Publ. Co., Amsterdam, pp. 1–385.

[7] Liszka, L. and Egeland, Alv.: 1964, 'Auroral Zone Ionospheric Research', Kiruna Geophysical Observatory of the Royal Swedish Academy of Science. Annual Summary Report No. AF 61 (052)-678.

[8] Liszka, L.: 1964, 'A New Method for determining the Altitude of Scintillation producing Irregularities', Kiruna Geophysical Observatory, Scientific Report No. 8; No. AF 61 (052)-678.

[9] Liszka, L.: 1964, 'An Investigation of the Height of Scintillation producing Irregularities', Kiruna Geophysical Observatory, Scientific Report No. 7; No. AF 61 (052)-678.

[10] Liszka, L.: 1964, 'A Study of the Scintillation-Rate Spectrum observed in the Auroral Zone', Kiruna Geophysical Observatory, No. AF 61 (052)-678.

OBSERVED LOW FREQUENCY FLUCTUATIONS IN SPACE

NORMAN F. NESS

*Extraterrestrial Physics Branch, Laboratory for Space Sciences,
NASA-Goddard Space Flight Center, Greenbelt, Md., U.S.A.*

Abstract. Measurements in the magnetosphere of low frequency magnetic fluctuations less than 1 KHz have been obtained recently with the OGO, ATS and IMP series of satellites. The OGO results indicate the existence of magnetically noisy regions of limited spatial extent within the magnetosphere. Fluctuations between 1Hz–1 kHz show two distinct classes of occurrence, steady and burst occurrences. The general character of the steady noise properties is in agreement with lower altitude ELF hiss measurements. The burst events may be associated with whistler loss cone instabilities. Transverse low frequency fluctuations with periods between 1 to 6 min have been observed on the ATS-1 satellite. Monochromatic oscillations of peak amplitude 2–20 gammas appear to represent the second harmonic of a standing Alfvén wave on a field line. Longer period oscillations of 6 or 8 min have also been observed and may represent an eigenmode oscillation of the magnetopause boundary. Compressional oscillations of the geomagnetic tail with periods of ∼ 20 min have been observed by Explorer-33 and probably represent the eigenmodes of the geomagnetic tail. A review of observations and theoretical studies of magnetic fluctuations below 1 kc is given in this paper.

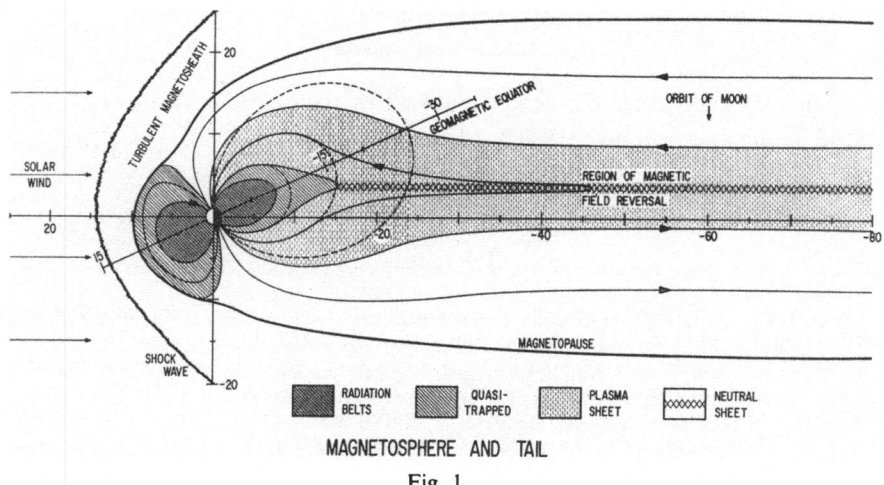

MAGNETOSPHERE AND TAIL

Fig. 1.

Short Bibliography on
Observed Low Frequency Fluctuations in Space

Anderson, K. A., Bincsak, J. H., and Fairfield, D. H.: 1968, 'Evidence for Hydromagnetic Waves of 3 to 5 min Period on the Magnetopause and Their Relation to Bow Shock Spikes', *J. Geophys. Res.* **73**, 2371.

Burlaga, L. F.: 1968, 'Micro-Scale Structures in the Interplanetary Medium', *Solar Phys.* **4**, 67–92.

Cummings, W. D., O'Sullivan, R. J., and Coleman, Jr., P. J.: 1968, 'Standing Alfvén Waves in the Magnetosphere', UCLA-IGPP Preprint No. 697.

Field, E. C. and Greifinger, C.: 1965, 'Transmission of Geomagnetic Micropulsations through the Ionosphere and Lower Exosphere', *J. Geophys. Res.* **70**, 4885–4889.

Greifinger, C. and Greifinger, P.: 1965, 'Transmission of Micropulsations through the Lower Ionosphere', *J. Geophys. Res.* **70**, 2217–2231.

Holzer, R. E., McLeod, M. G., and Smith, E. J.: 1966, 'Preliminary Results from the OGO-1 Search Coil Magnetometer: Boundary Positions and Magnetic Noise Spectra', *J. Geophys. Res.* **71**, 1481–1486.

Holzer, R. E., McLeod, M. G., Olson, J. V., and Russell, C. T.: 1968, 'The Magnetic Component of Plasma Waves associated with the Interaction of the Magnetosphere and Solar Wind', presented at International Meeting on Physics of the Magnetosphere, Washington, D.C.

Judge, D. L. and Coleman, P. J.: 1962, 'Observations of Low Frequency Hydromagnetic Waves in the Distant Geomagnetic Field: Explorer-6', *J. Geophys. Res.* **67**, 5071–5090.

Lerche, I.: 1966, 'Validity of the Hydromagnetic Approach in discussing Instability of the Magnetosphere Boundary', *J. Geophys. Res.* **71**, 2365.

McClay, J. F. and Radoski, H. R.: 1967, 'Hydromagnetic Propagation in a Theta Model Geomagnetic Tail', *J. Geophys. Res.* **72**, 4525–4528.

McDonald, G. J. F.: 1961, 'Spectrum of Hydromagnetic Waves in the Exosphere', *J. Geophys. Res.* **66**, 3639–3671.

Ness, N. F.: 1968, 'The Geomagnetic Tail', 1969, *Rev. Geophys.* **7**, 97–127.

Patel, V. L.: 1968, 'Magnetosphere Tail as a Hydromagnetic Waveguide', *Phys. Letters* **26A**, 596–597.

Russell, C. T., Holzer, R. E., and Smith, C. J.: 1968, 'OGO-3 Observations of ELF Noise in the Magnetosphere: Part 1. Spatial Extent and Frequency of Occurrence', UCLA-IGPP Preprint No. 696

Sari, J. M. and Ness, N. F.: 1968, 'Power Spectra of the Interplanetary Magnetic Field', NASA-GSFC Preprint X-616-68-318.

Siscoe, G. L., Davis, Jr., L., Smith, E. J., Coleman, Jr., P. J., and Jones, D. E.: 1967, 'Magnetic Fluctuations in the Magnetosheath, Mariner-4', *J. Geophys. Res.* **72**, 1–17.

Smith, E. J., Holzer, R. E., McLeod, M. C., and Russell, C. T.: 1967, 'Magnetic Noise in the Magnetosheath in the Frequency Range 3–300 Hz', *J. Geophys. Res.* **72**, 4803–4813.

Zmuda, A. J., Martin, J. H., and Heuring, F. T.: 1966, 'Transverse Magnetic Disturbances at 1100 km in the Auroral Region', *J. Geophys. Res.* **71**, 5033–5045.

BARIUM RELEASE EXPERIMENTS NEAR
THE MAGNETIC EQUATOR AT THUMBA, INDIA

E. RIEGER

Max-Planck-Institut für extraterrestrische Physik, Garching near Munich, Germany

Abstract. On March 28, 30 and 31, 1968 four Nike-Apache rockets carrying Ba-release experiments were launched during twilight to investigate the electric field in the upper atmosphere at the geomagnetic equator in a height range of 150–250 km. The ionospheric electric field was derived from the observation of the movements of the artificial ion clouds. The drift of the ion clouds was to the West for the post-sunset experiments with the East–West component ranging from 50 to 150 msec^{-1} and upward with a velocity of about 1/3 of the horizontal component. This gives an electric field of 1.8 to 5.4 Vkm^{-1} upward and about 0.6 to 1.8 Vkm^{-1} eastward.

For the pre-sunrise experiment the velocity of the ion clouds was directed eastward and downward and the electric field pointing downward and westward was of the same order of magnitude as for the post-sunset experiments.

D'Angelo (ed.), Low-Frequency Waves and Irregularities in the Ionosphere. All rights reserved